# Keeping Up with Technologies to Improve Places

# Keeping Up with Technologies to Improve Places

Edited by

Eva Vaništa Lazarević,
Milena Vukmirović,
Aleksandra Krstić-Furundžić
and Aleksandra Đukić

Cambridge
Scholars
Publishing

Keeping Up with Technologies to Improve Places

Edited by Eva Vaništa Lazarević, Milena Vukmirović, Aleksandra Krstić-Furundžić and Aleksandra Đukić

This book first published 2015

Cambridge Scholars Publishing

Lady Stephenson Library, Newcastle upon Tyne, NE6 2PA, UK

British Library Cataloguing in Publication Data
A catalogue record for this book is available from the British Library



ISBN (10): 1-4438-7739-5
ISBN (13): 978-1-4438-7739-8

# TABLE OF CONTENTS

Preface ..... viii

**PART 1: URBANISM AND TECHNOLOGIES**

Chapter One ..... 2
Overcoming Barriers to Growth
Stephen Platt

Chapter Two ..... 14
Saving Urban Planning from another Utopian Model
Danijela Milojkic and Marija Maruna

Chapter Three ..... 24
Place Competitiveness Expressed through Digital Data
Milena Vukmirovic and Eva Vanista Lazarevic

Chapter Four ..... 39
Transparency of Scale: Geographical Information Program (Google Earth) and the View from Beyond
Pavle Stamenovic, Dunja Predic and Davor Eres

Chapter Five ..... 48
Retail: New Technologies and Urban Centrality
Martin Brabant

Chapter Six ..... 62
Comfort of Open Public Spaces: Case Study New Belgrade
Aleksandra Djukic, Nevena Novakovic and Srđan Stanković

Chapter Seven ..... 79
Vertical Public Spaces
Sorana Radulescu

Chapter Eight ........................................................................................ 93
Contemporary Interventions in Historic Places:
The Example of Thessaloniki Metro
Stavros Apostos

Chapter Nine........................................................................................ 102
Spatial and Social Aspects of an Arsenal's Transformation: Military Port in Tivat into Nautical-Tourism Settlement and Port "Porto Montenegro"
Goran Radovic

Chapter Ten ........................................................................................ 115
An Overview of Urban Regeneration Projects in Turkey
Mehmet Cete and Yunus Konbul

**PART 2: ARCHITECTURE AND TECHNOLOGIES**

Chapter Eleven ........................................................................................ 126
Integrated Architectural Complexity: From Abstraction to Technology and Materialization
Rada Cahtarevic, Dzenana Bijedic and Amra Taso

Chapter Twelve ........................................................................................ 136
The Interplay of Music and Architecture: Layering of Sound and Space
Anja Kostanjsak and Morana Pap

Chapter Thirteen........................................................................................ 146
Evolution Digitalized: Architecture of Sublime Dream
Mihailo Popovic and Vladimir Milenkovic

Chapter Fourteen ........................................................................................ 155
Between the Place and Non-Place: Architecture and Territory on the Example of Skopje
Sasa Tasic, Mitko Hadzi Pulja and Minas Bakalcev

Chapter Fifteen ........................................................................................ 165
Environment Certification of Rehabilitation Design Projects:
PUT and SHU Buildings as Case Studies
Florian Nepravishta, Gerta Veliu and Ramadan Alushaj

Chapter Sixteen ........................................................................ 178
Green Urban Strategies in Thessaloniki in the Context of Crisis
Evangelia Athanassiou

Chapter Seventeen .................................................................... 189
Influence of Glass Component Joints on the Structural Glass Façade Design
Aleksandra Kristic-Furundzic, Tatjana Kosic and Jefto Terzovic

Chapter Eighteen ...................................................................... 209
The Influence of New Technologies on Modern City Facades
Jasna Cikic-Tovarovic, Jelena Ivanovic-Sekularac and Nenad Sekularac

Chapter Nineteen ...................................................................... 218
New Composite Slab System: Lightweight Concrete, Steel Sheeting and Reinforcement
Zoran Šobić, Jelena Milošević and Miodrag Nestorović

**PART 3: PLACES, TECHNOLOGIES AND RELATED FIELDS**

Chapter Twenty ........................................................................ 230
The Role of Cartography in Making a "Smart City": Case Study of Indjija
Dragutin Protic, Ivan Vucetic and Ivan Nestorov

Chapter Twenty One.................................................................. 238
Re-Thinking Infrastructure Project for Metropolis: Laboratory Granada
Juan Luis Rivas Navarro and Belen Bravo Rodriguez

Chapter Twenty Two ................................................................. 248
Personal Rapid Transit: A Sustainable Urban Transport System
Ljupko Simunovic, Luka Novacko and Mario Cosic

Chapter Twenty Three ............................................................... 259
Flightpath to an Environmental Friendly Air Transport
Ivana Cavka and Olja Cokorilo

# PREFACE

*Keeping up with technologies to improve places* is a book of the selected best papers presented at the 1st International Academic Conference on Places and Technologies, held at the University of Belgrade – Faculty of Architecture on 3rd and 4th April 2014. The conference was organized by three partner organisations; University of Belgrade - Faculty of Architecture, Professional Association Urban Laboratory and University of Belgrade - Faculty of Philosophy, with the aim of bringing together leading researchers, professors and PhD students, as well as practitioners, in order to create a platform for sharing knowledge and know how in the fields of growth, new technologies and environment. The stated goals point to the necessity of a multidisciplinary approach, identifying relationships between technological development, environmental protection and social change. Consequently, the conference program and speakers were to focus on the knowledge of several academic disciplines: engineering and technical sciences, humanities and social sciences. The main tasks of the conference are defined as follows; discussion on the current issues related to the future of society and places, design of places, facilities and infrastructure in line with new and future needs of inhabitants, development of institutions and regulations with the aim of creating an appropriate and high quality environment, and the creation of favourable conditions for the advancement of innovation and business.

As we have received a variety of very interesting and innovative research papers as made by young scientists and their mentors from all over Europe, we are proud to share them all in one place, something which was made possible by Cambridge Scholars Publishing, who recognized the importance of the subject and the value of the papers, in deciding to publish this book. In accordance with the main goal and tasks, the book includes 25 papers structured into three main parts. The first part, *Urbanism and Technologies*, covers the topics of a) urban planning and technologies, b) big data, apps, social networks and micro blogs in urban planning and design, c) urban design and technologies and d) historical centres, building heritage and technologies. The second part, *Architecture and Technologies,* includes papers in the fields of a) sustainability, sustainable buildings and technologies, b) green strategies and technologies,

c) innovative materials, systems and technologies and d) cultural patterns, architecture and technologies. The third part, *Places, technologies and Related fields,* deals with the issues of a) geodesy and modern cartography and b) mobility and technologies.

As the editors of this book, we want to express our appreciation to all the authors and our partners from Cambridge Scholars Publishing for their kind invitation, cooperation and help in preparation of the book.

Editors

Belgrade,

25th December 2014

# PART 1:

# URBANISM AND TECHNOLOGIES

# CHAPTER ONE

# OVERCOMING BARRIERS TO GROWTH

STEPHEN PLATT[1]

**Abstract** | Cambridge is the centre of the greatest concentration of high technology firms in Europe. Economic activity in the sub-region has expanded over the last 50 years and there is intense and increasing pressure for further growth. But prior to the mid-nineties growth was highly constrained by policies dating back to the 1950s. (Holford and Wright, 1950)

In 1997, concerned about the constraints on growth in Cambridge, the University Department of Architecture brought together people from local government, planning, development and the business community to discuss the future of the region in a forum called Cambridge Futures. The Cambridge Futures initiative, and the public consultation, had a major impact on development in the Cambridge area. In particular it overcame local politicians' perceptions that the public were opposed to growth and demonstrated that it was possible to engage a wide range of stakeholders in thinking strategically about the future.

This paper describes the results of the modelling exercise and findings of the public consultation. It also analyses the factors in Cambridge Futures' success: leadership, ambition, evidence, participation and cooperation. This stakeholder and public engagement is relevant to other cities facing public opposition to change and the analysis of key success factors is relevant to other places trying to overcome barriers to high tech economic development.

**Keywords** | Growth, barriers, Cambridge, futures, planning

---

[1] Chairman, Cambridge Architectural Research Ltd. www.carltd.com

## Introduction

This paper describes the period from 1997-2004 when the policy of restricting the growth of Cambridge was ditched and the growth agenda embraced, and it brings the story up to date by describing how the city is dealing with the challenges of the National Planning Policy Framework. Cambridge has successfully developed a process of gaining public backing and community support for change, by setting up organisations to deal with coordinating the big issues and by developing and delivering a strategic plan for growth. This is the story of how people in the University, local authorities and business came together to debate the issues, how a case for change was made, and how the planners and elected members implemented the changes. (Platt, 2013)

## Context

One of the UK Government's priorities is to support growth, and to make sure that the planning system plays its part in this. But growth is not always easy since people tend to oppose major changes and because we need to balance growth with environmental and social concerns. The Government has addressed this by simplifying the planning system with the new National Planning Policy Framework (NPPF, 2012). The Cambridge sub-region is a prime example of the dilemma of balancing growth and restraint. The population is forecast to increase by 28% and the number of households by 33% by 2031.

## Cambridge Phenomenon

Cambridge is the centre of the greatest concentration of high technology firms in Europe. Economic activity in the area has expanded over the last 50 years and there is intense and increasing pressure for growth. (Wicksteed, 1985; Garnsey and Mohr, 2010) A new book entitled The Cambridge Phenomenon (Kirk and Cotton, 2012) identifies 1,000 technology and biotechnology companies in the Cambridge cluster, plus 400 support organisations, that together employ over 40,000 staff. Robert Koepp (2002) nominated Cambridge as Europe's silicon fen and compared it to Silicon Valley, California. He describes the evolution of the symbiotic relationship between University science and engineering and high tech innovation. He also analyses the logistics of location. Segal Quince Wicksteed, in 2011, said *"planning of future development is a hot topic in the Cambridge area where the challenge is to provide for rapid development in sustainable ways that maintain quality of life*".

## Cambridge planning history

The history of postwar planning in Cambridge since 1950 can be divided into five distinct periods. (See Appendix 1)

1 Restricting growth in the city 1950-65
2 High-tech expansion of the University 1966-79
3 Neighbourhood planning and green-belt policy 1980-98
4 Housing growth agenda 1997-2009
5 Plan implementation 2010 to date

Prior to 1996, planning policies sought to restrict the population of Cambridge and constrain development to within the green belt and it was not until the mid 1990s that things began to change. The City Council produced draft proposals in the 1997 Local Plan for an urban extension but the proposed changes to the green belt did not get through the planning approvals process. The University was also having difficulty getting approval for proposals. The real watershed was a visit by a delegation of the Malaysian Government in 1997 which outlined their plans for a super high-tech corridor. Discussions identified two apparently conflicting goals: Cambridge must exploit its position to become a global player in high-tech and Cambridge and environs must preserve its architectural and environmental heritage. (Ablett et al, 1998) A report by Mott (1969) identified ICT as key in resolving this dilemma and made nine recommendations about how ICT might be applied to improve various sectors and systems, including transport and traffic management. But in 1998 the City Council produced an urban capacity study which concluded that although there might be scope for some additional capacity it was considered to be minimal. (Cambridge City Council, 1998) The consequences of these planning polices may be summarized as follows:

> *The increase in the number of jobs and households within a restricted land supply had led to rising property prices. People employed in the City and its fringe had been forced to live beyond the Green Belt where cheaper housing more than offsets the cost of travel into Cambridge.*
>
> *Population growth in surrounding villages and market towns had been amongst the highest in the country. As a result there was a daily influx of nearly 40,000 workers from outside the City, increasingly outnumbering resident workers. Congestion in the access roads had risen, increasing emissions and pollution.*

## Cambridge Futures

In 1997, concerned about the constraints on growth in Cambridge, Louis McCagg, a volunteer in the University's development office, persuaded academics in the University Department of Architecture to bring together people from local government, planning, development and business in a forum called Cambridge Futures. (Carolin, 2004)

| | Economic Efficiency | Social Equity | Environmental Quality |
|---|---|---|---|
| 1. Minimum Growth | ● | ● | ●●● |
| 2. Densification | ●●●●● | ●●●●● | ● |
| 3. Necklace | ●● | ● | ●●● |
| 4. Green Swap | ●●●● | ●●● | ●● |
| 5. Transport Links | ●●● | ●●● | ●●●● |
| 6. Virtual Highway | ●● | ●● | ●●●●● |
| 7. New Town | ● | ● | ●●●● |

*Qualitative comparison of the seven options – the more bullets, the better the option.*

*Figure 1: Cambridge Futures – comparison of options*

Cambridge Futures produced the evidence and modelled options that were influential in changing attitudes to growth. The consequences of each option were assessed in terms of economic efficiency, social equity and environmental impact. Virtual reality movie clips were made to help the public visualise what each option might look like. (Echenique, 1999)

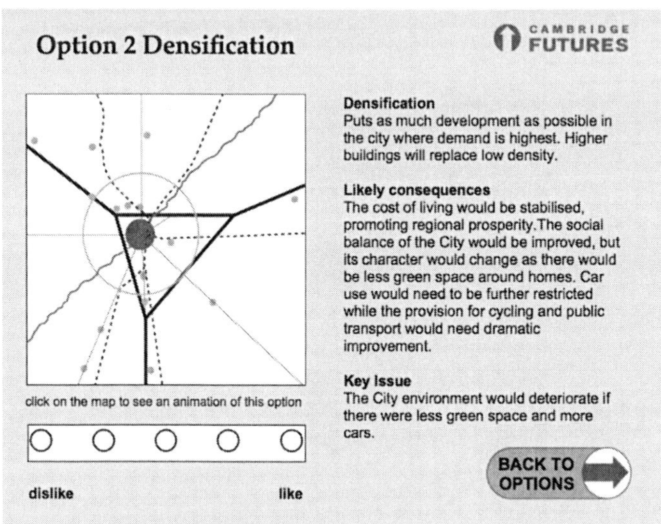


*Figure 2: Cambridge Futures–example of an options survey of public opinion*

In 1999 Cambridge Architectural Research was tasked with surveying the public's reaction to these options and they were responsible for devising and administering a survey that formed part of an exhibition at various venues in the region. (Platt, 1999a) People were able to comment on the proposals by filling in a tear-off or by using a computer-based interactive questionnaire. A total of 650 people answered the survey. 52% lived in the city and 48% in the region. 59% used the interactive computer survey. The options were described in terms people could relate to. Respondents were asked to indicate on a simple 5-point scale how much they liked or disliked each option. There was also a set of eight controversial statements, for example the headline in the local press at that time – "Cambridge is full".

The findings of the survey were published in a clear report on two sides of an A3 sheet. This simple document proved to be influential. Professor Peter Carolin said at the time: *The public consultation exercise was a most amazing piece of research, it proved to politicians that Joe Public is not as stupid as the tabloid press would have us believe.* (Platt, 1999b)

One of the main findings was that there was little support for the 'status quo'. Options that won the least support were closest to the then current planning policies of 'minimum growth' for the City of Cambridge and 'necklace development' of villages outside the green belt. And people accepted that Cambridge could not be kept the same and that if nothing was done it would only get worse. There was also clear backing for public transport and reducing traffic congestion, which led to Cambridge Futures 2, for which CAR also did the public consultation.

Not surprisingly, there was also clear evidence of nimbyism, and people wanting the growth somewhere other than where they were living. Most importantly, there was clear support for growth.

The popular view was that people wanted a balance of development in Cambridge and in the region. The survey suggested that a planning strategy which aimed for some growth in Cambridge through densification and selective expansion into the green belt, together with growth outside Cambridge based on public transport links, would be most likely to meet the aspirations of people in the region. And this is more or less what the Structure Plan delivered four years later and what we are experiencing now.

About the same time, in 1998, the Cambridge Network was founded by an influential group comprising the then Vice-Chancellor of the University of Cambridge, Lord Broers, with businessmen and entrepreneurs. Peter

Dawe, an entrepreneur and member of the Network, argued that the supply of land for housing must be increased, that the environment needed enhancing and transport needed improving. (Dawe and Lindsay, 2001)

## The growth agenda

In parallel with the work of Cambridge Futures the local authorities developed a radical growth strategy for the Cambridge Sub-region through the review of Regional Planning Guidance for East Anglia which was adopted in 2000. (RPG6, 2000) This set an ambitious target for housing growth together with a sequential approach to its location from within the built up area of Cambridge outwards. In 2000 the Government issued technical guidance on doing multi-modal transport studies. (DETR, 2000) The result was the Cambridge to Huntingdon Multi-Modal Study which provided the policy basis for the Cambridge to St Ives guided busway and the planned improvements to the A14 corridor. (Department for Transport, 2003)

Within the strategic framework set by RPG6 the local authorities commissioned studies by Colin Buchanan (2001), DEGW and Llewelyn-Davies that informed the Structure Plan adopted in 2003. (Cambridgeshire County Council, 2003) This made provision in Cambridgeshire and Peterborough for 70,200 additional homes between 1999-2016. The plan maintained the green belt as a key mechanism for defining the limits of urban growth and preserving the character of Cambridge but also detailed the location and phasing of land to be released from the green belt. The Structure Plan was accompanied by a Local Transport Plan. (Smith, 2003)

The strategy was carried forward into the Regional Spatial Strategy (RSS) for the East of England. This set an ambitious target for growth of 2,800 dwellings per year, a 40% increase in the rate achieved during the 1990s. In 2004 Cambridge was included in the Government's Sustainable Communities Plan as a key component of the London-Stansted-Cambridge-Peterborough Growth Area.

Both the Cambridge City and South Cambridgeshire District Councils produced Local Plans. (Cambridge City Council, 2006) In 2008 Central Government issued the East of England Plan as a revision to the RSS. This upped the numbers of jobs and houses. In the Local Plan Review 2013 the City has been consulting on Issues and Options to try to meet present challenges.

The recent abolition of the RSSs, together with the winding up of Regional Development Agencies and other coordinating bodies such as Cambridgeshire Horizons has meant that the local authorities are now reliant on a 'duty to cooperate' across borders to maintain the vision for strategic planning in the Cambridge sub-region.

In 2013 the Planning Advisory Service (PAS), part of the Department of Communities and Local Government, commissioned the author of this paper to conduct a case study of planning in the Cambridge area and how the policy of restricting growth of the city was ditched and the growth agenda embraced.

The study reported on interviews with 13 key players in the drama. They included planning officers and planning portfolio-holding elected councillors who were in post when the new growth strategy was being debated and decided. They also include the leading lights in Cambridge Futures. The results of these interviews are summarized under five headings: leadership, ambition, evidence, participation and cooperation.

## Leadership

Leadership is critical in the process. Planners and councillors in the City and County were crucial in leading and delivering the agenda for change. Along with others on the Executive of Cambridge Futures, lead players provided a mix of personalities, background and expertise. Timing was also crucial. Cambridge Futures started just before the Structure Plan process and influenced the system. Later and it would have been too late.

## Ambition

The Cambridge growth strategy adopted through this process represented a step change in thinking about the future shape of Cambridge and its surrounding sub-region. Cambridge Futures drove the ambition and helped shape the debate that guided the first three stages of the process by articulating the consequences of different approaches to the growth challenge, and by illustrating the options in a way that stimulated wide public engagement. It also took some of the flak, clarified the degree of opposition to growth, and was useful in raising the profile of all that officers and members were trying to achieve.

## Evidence

Clear and transparent evidence was crucial in informing decision-making and in convincing people about growth. Cambridge Futures modelled 7 options, including no change. The outcomes of each were summarised in terms of things people could relate to such as jobs, traffic, house prices and impact on open space.

## Participation

Cambridge Futures asked people in the Cambridge sub-region their opinion about the options and whether they agreed with a set of statements. The option that won least favour was the status quo. People recognised that if nothing was done, things would get worse. The statement that received least agreement was 'Cambridge is full'. This gave elected members the confidence that they had public support for a growth strategy. This meant that Structure Plan proposals about the green belt, densification and new settlement did not meet with the kind of opposition other places have experienced when proposing big changes.

## Cooperation

The final thing to stress is the importance of cooperation in a place that lacks a unitary authority. Prior to 1996 the various local authorities – Cambridge City, South Cambs, East Cambs and Huntingdon District Council, Peterborough City Council and Cambridgeshire County Council – did not cooperate or coordinate their planning. Partly thanks to Cambridge Futures providing a forum for officers and members to meet on neutral territory, and perhaps more importantly because the four key local authority leaders of the process came into post at the time or two or three years later, things changed in the late nineties and it became possible to build a new consensus around taking a more positive approach to accommodating future growth.

## Applying the lessons elsewhere

Can these insights from Cambridge be applied elsewhere? Or are the timing and circumstance in the late nineties so specific that Cambridge is a unique case? To make progress the key lesson is to bring opinion formers together in a neutral forum to discuss the issues in a non-adversarial manner.

We are fortunate in the economic strength in the Cambridge region. Cambridge is a vibrant area with a high quality of life. People want to move into the area and firms want to locate here. It is clearly a bonus planning in a university town with a strong social fabric. Cambridge Futures was also fortunate with the timing. Whether the Cambridge experience is replicable depends on whether people are willing to work together, to find the leaders, develop the ambition, produce the technical unbiased evidence and win over hearts and minds to the big idea.

In support of applying the lessons elsewhere the author applied a similar approach to Cambridge Futures, albeit in a more limited way, in three other cities – Aylesbury, Maidenhead and Medway. The findings again demonstrated that the general public is much less opposed to growth than politicians suppose if presented with good evidence and a set of options. (Platt et al., 2004)

## Key lessons

The first key lesson is the importance of having an agreed strategic vision amongst the key players, both within the various local authorities, but also within the business community, academia and the wider community.

A coherent structure of strategic planning across an economic sub-region to make decisions about the location of growth, the enhancement of the landscape and the provision of physical, social and cultural infrastructure is also important.

The value of a cross-sector think tank such as Cambridge Futures is that it works alongside the statutory authorities and can 'think the unthinkable' at an early stage of a plan review process. It can also engage local communities in a more open and accessible way than the more formal statutory processes. This think tank produced the comprehensive evidence base for testing and evaluating different strategic options for growth.

Finally and not least, local leadership, both from political leaders and from Planning Directors, together with maintaining continuity of leadership over time is crucial. Cambridge also benefitted from a dedicated delivery team, Cambridgeshire Horizons, which supported local authorities in implementing their growth strategy and helped them define and secure high quality standards in the new communities being planned.

## Conclusions

To date only part of the vision for Cambridge has been achieved. Densification of the City and some of the urban extensions are happening, but two of the biggest pieces in the jigsaw – huge missing segment of the Northern Fringe East and the new settlement of Northstowe – will not be delivered by 2016. On transport, there are better rail links to London and Felixstowe, the guided busway, and new stations, but the A14 upgrading and a traffic demand management system are still pending. There is clearly a great deal still to be done.

## References

Ablett, S., Broers, A. et al (1998) Cambridge 2020: Meeting the challenge of growth. Analysys Publications

Buchanan, C (2001) Cambridge Sub-Region Study Colin Buchanan and Partners in association with Bone Wells Associates and Wardell Armstrong. Draft Final report February 2001

Cambridge City Council (1998) Cambridge: an estimate of urban capacity.

Cambridge City Annual Demographic and Socio-Economic Report, April 2011. Cambridgeshire County Council Research Group.

Echenique, M (1999) Cambridge Futures, Cambridge University Press, Cambridge 48pp

Cambridge City Council (2006) Cambridge Local Plan.

Cambridgeshire County Council (2003) Cambridgeshire & Peterborough Structure Plan: Planning for success

Carolin, P (2004) Cambridge University City, Cambridge Futures: enabling consensus on growth and change

Dawe, P and Lindsay, W (2001) In our back yard: a vision for a small city. Published by Dawe Consulting Ltd.

Department of Environment, Transport and the Regions Guidance on the Methodology for Multi-Modal Studies Volumes 1 and 2. DETR, London, 2000.

East of England Plan (2008) The revision to the Regional Spatial Strategy, Government Office for the East of England, CL

Kirk, K and Cotton C (2012) The Cambridge Phenomenon: 50 Years of Innovation and Enterprise, Third Millennium Information

Koepp, R (2002) Clusters of Creativity: Enduring Lessons on Innovation and Entrepreneurship from Silicon Valley and Europe's Silicon Fen. John Wiley

Garnsey, E and Mohr, V (2010) Technology Cluster: Evidence from Cambridge, UK

Holford, W and Wright, H M (1950) Cambridge Planning Proposals. A report to the Town and Country Planning Committee of the Cambridgeshire County Council

Mott report (1969) Report on Relations between the University and University of Cambridge, Cambridge University Reporter, 22 October.

National Planning Policy Framework (2012) Department for Communities and Local Government

Department for Transport (2003) Multi-Modal Studies: Introduction to the Guidance on the Methodology for Multi Modal Studies. Transport Analysis Guidance (TAG) Unit 1.2.1 June 2003

Platt, S (1999a) Cambridge Futures Survey Report. Cambridge Architectural Research, October

Platt, S (1999b) Cambridge Futures: Partnership in planning in the economic landscape. Cambridge Architecture 43 Winter 1999 and in Locum Destination Review, July 2000

Platt, S, Fawcett, W and de Carteret, R (2004) Public Views of Development Options in the South East, Joseph Rowntree Foundation. http://www.jrf.org.uk/publications/public-views-development-options-south-east

Platt, S (2013) Backing for a big idea: Consensus building for strategic planning. Planning Advisory Service. DCLG

RPG6 (2000) Regional Planning Guidance for East Anglia to 2016. Regional Planning Guidance Note 6. Government Office for the East of England. Department of the Environment, Transport and the Regions. November 2000

Segal Quince Wicksteed (2011) Cambridge Cluster at 50: The Cambridge economy: retrospect and prospect. Final report to EEDA and partners March 2011

Smith, B (2003) Cambridgeshire Local Transport Plan 2004-2011 Moving transport forward. Cambridgeshire County Council, July 2003

South Cambridgeshire Local Plan. Adopted February 2004. South Cambridgeshire District Council

Studdert, P (2013) Design and Distinctiveness in the Cambridge Growth Area, Rudi.net. http://www.rudi.net/books/12421

Wicksteed Segal Quince (1985) Cambridge Phenomenon: The Growth of High Technology Industry in a University Town. Segal Quince & Partners

# Appendix 1

**Cambridge Planning History**

*Period of restricting growth in the city and directing growth to surrounding settlements*

1950 Cambridge Planning Proposals, Cambridgeshire CC, Holford W & Wright M
1952 Cambridgeshire County Development Plan, Cambridgeshire CC
1955 Town Map, Cambridgeshire County Council
1956 A Guide to the Cambridge Plan, Cambridgeshire County Council, Derek Senior
1957 Green Belt defined, approved in principle by minister, and given status of planning policy
1964 Proposal for Bar Hill approved
1964 South East Study, Central Government, Reference to growth pressures in Cambridge
1965 Cambridgeshire Development Plan Review, Cambridgeshire County Council
1965 Cambridge Town Map, Cambridgeshire County Council, Adopted

*Period of high tech expansion of the University*

1966 The Future Shape of Cambridge, Cambridge City Council, Gordon Logie,
1968 East Anglia – A Study, East Anglia city region. Economic Planning Council
1969 Mott Report, Cambridge University, Proposes limited research related expansion
1969 County Development Plan, Cambridgeshire County Council, Excludes Cambridge
1970 Cambridge Science Park
1971 Cambridge Study Area Review, Cambridgeshire County Council, Adopted
1974 A Study of the Cambridge Sub Region: Professor John Parry Lewis DoE
1974 Swinnerton-Dyer Report, University of Cambridge, Long term development
1979 Cambridgeshire Structure Plan, Cambridgeshire County Council

*Period of neighbourhood planning and continuation of green belt policy*

1984 Newnham and West Cambridge District Plan, Cambridge City Council
1985 The Cambridge Phenomenon, Segal Quince Wicksteed
1986 Romsey Local Plan, Cambridge City Council, ,
1988 St Matthews Local Plan, Cambridge City Council
1989 Cambridgeshire Structure Plan, Cambridgeshire County Council
1991 Regional Planning Guidance for East Anglia
1992 Cambridge Green Belt Local Plan, City, County, SCDC
1996 Cambridge Local Plan, Cambridge City Council
1998 Cambridge Green Belt Towards 2016, Cambridge City Council

*Period of housing growth agenda*

1997 Cambridge Futures founded
1999 Cambridge Futures 1, Land use study and consultation
2000 Regional Planning Guidance for East Anglia to 2016 (RPG6), Government Office
2003 Cambridgeshire & Peterborough Structure Plan, Cambridgeshire CC
2004 Cambridgeshire Horizons established
2004 Cambridge Futures 2, Transport study and consultation
2006 Cambridge Local Plan, Cambridge City Council
2008 Regional Spatial Strategy for the East of England, Government Office
2007 Lessons from Cambourne, CAR

*Period of plan implemenation*

2011 Cambridgeshire Quality Panel set up to review major development proposals
2012 NPPF, National Planning Policy Framework promotes localism
2013 Regional Spatial Strategy for East of England scrapped
2013 Development in CB1 around Cambridge railway station
2014 Urban extension in Trumpington Meadows/Clay Farm, SW Cambridge
2015 Urban extensions in NW Cambridge begin on site

# CHAPTER TWO

# SAVING URBAN PLANNING FROM ANOTHER UTOPIAN MODEL

DANIJELA MILOJKIC[1] AND MARIJA MARUNA[2]

**Abstract** | Criticism of wide civic participation, as a means for every citizen to actively engage in decision making, including planning processes, has a long history and numerous voices. Although the model is plausible for its openness and somewhat necessary for achievement of contemporary democratic standards (governance transparency, accountability etc.), civic participation has long been criticized for its ineffectiveness, inefficiency and utopian characteristics. Yet, the very notion of the need for the wide public to participate in important decision making processes came along with another novelty—the development and the widespread use of informational technologies. Emerging genres of online platforms combine specific technologies with various actors and activities, thus fostering different types of interpersonal interactions. They are becoming sources of information and platforms for public expression, and complementing traditional media. It is inarguable that new technologies, which support deliberation and information dissemination, can contribute to the increase of civic participation. A completely new range of modes of expression enables citizens to voice their opinions and to be heard, while fast and novel forms of information dissemination contribute to citizens' awareness and knowledge about different matters. It may seem that the circle of conditions for successful civic participation has now been closed. In this article we will discuss the possibilities of civic participation/engagement in urban planning processes through

[1] Head of group for the local economic development, City municipality of Mladenovac, st.milojkic@gmail.com
[2] Associate Professor, University of Belgrade-Faculty of Architecture, Belgrade, Serbia, m.ma@sezampro.rs

informational technologies, in relation to the conditions that civic participation needs and the critiques it is exposed to.

**Keywords** | Urban planning process, Civic participation, Online communication tools, Information gathering

## Civic participation: problems with the masses, problems with individuals

The participatory model in urban planning was derived from "efforts to reinterpret a progressive meaning for democracy in Western societies" (Healey, 1992, 145). While considering progressive democracy, Healey aligned herself with the position of those authors who state that democracy needs a system which promotes discussion, debate and competition among many divergent views, and that open debate, access to power centres and general political participation are key requirements for democratic public life. In an era of what are often referred to as democracy crusades across the international scene, participative models of urban planning (among others) play roles in generating transparency and accountability of governance. When successful, they can contribute to the quality of proposed plans and strategies by introducing different views and sources of information. Still, the main problem with participatory planning lies in its applicability in different situations and consequently in its success rate. It became evident that on many occasions participatory processes do not function, and that in some cases they can be counterproductive (Bohman, 1966; Warren, 1992; Mansbridge, 1999; Guthman and Thompson, 1996). This deficiency with participation was adopted by planning along with the benefits of the participatory model in democracy, a model which has been subjected to considerable criticism. The critique is mostly focused on this model's utopian character, depicted in what Hauptmann (2001, 399) calls "nostalgia for simpler societies and majoritarian biases"—the presumption that society is willing and apt for participatory processes, that it has the necessary characteristics to support the open and democratic participatory decision-making processes, and that the decisions made by citizens would be better, fairer and impartial in relation to those made by experts (Healey, 1992; Innes, 1995; Fischer and Forester, 1993).

Deliberative democrats who expose the participatory model to critique literally number the characteristics of society which disqualify it for participatory processes and point to problems with the demands of participatory processes with regard to citizens. They do not argue against participation in general, but envision it as less of a "mass" event—as an engagement of groups and organizations with clear and operable interests and goals. They claim that citizens can contribute to politics meaningfully and intelligently and should be enabled to do so more often if they want to; but it must not be an obligation for them, or something that governance and the entire political system rely and depend on (Cohen, 1997). The advantage of deliberative democracy (in the sense of a system with a less utopian character) is that it does not involve an idea of the ideal society where everyone's opinion is equally relevant. They suggest a system where opinions clash and the one that passes the test of the public can be accepted. They disagree with the notion that citizens' participation in decision making leads to changes in their own interests and changes in the social structure itself, as alleged by participatory theorists (Hauptman 2001).

The first set of critiques directed at participatory democrats concerns citizens' **will to participate** in the first place. Warren (1999a; b; 1992; 1996) based his criticism on understanding that the complexity of society makes the ideals on which participative theory relies impossible to exercise. According to him, the theory of participatory democracy is essentially too reliant on the mistaken belief that people enjoy politics. Warren believes that this is a romantic dogma. Bohman (1996) argues that participatory democrats extrapolated participation in every field of life and cast a utopian shadow on their own positive democratic ideals. This requires participants to fulfil excessive obligations, as well as hold excessive qualifications, while the processes themselves take forever. According to Warren, citizens will see participation in politics as a burden, without great results, and let themselves succumb to cynical apathy, leaving the entire process to a handful of enthusiasts who will make decisions on behalf of others on the authority that this participatory process will give to them (ironically in relation to the ideals of participatory theory). Also, for most people participation in decision-making activities means loss of personal freedom (in the sense of free time) without specific gain (Guthmann, 1993).

Another issue is the problem of **majority strength and the psychology of the masses**. Benhabib (1996) and Gutmann and Thompson (1996) claim that most radical versions of participatory democracy are not apt to protect individual interests and human rights from the tyranny of the majority and mandatory consensus. In their influential work, Dye and Zeigler (2008) state that the irony of democracy is that the elite must rule wisely if the "rule of the people" is to survive. In their view, the masses are "authoritarian, intolerant, anti-intellectual, nativist, alienated, hateful and violent." Gutmann and Thompson (1996) state that in order for the political discussion to be of good quality, it has to be carried out among citizens of equal status who can provide each other with reasonable, carefully constructed and morally justifiable arguments in the context of mutual respect. Ethically speaking, it is not certain that decision-making by all citizens results in a high quality of laws and policies, nor that mere participation is a justifiable reason for such outcomes.

Lastly, the very idea of wide participation undermines one of participation's key concepts: that it can help **self-development of personality**. Bachrach and Botwinick (1992) and Warren (1992) state that this is based on the mistaken assumption that ordinary citizens have the capacity to expand their interests in order to be able to identify the common good in those interests and to devote themselves to it. The utopian component is contained in the premise that implementation requires a society where everyone is equal and a society which is not dominated by greed. According to these authors, we should not have a lot of faith in the reasonableness of ordinary citizens, and should consider the masses to be largely passive on matters of policy and poorly informed about public affairs and politics.

The way in which participatory processes and their participants are envisioned hampers their efforts to achieve transparency of processes, quality of developed planning documents and accountability of decision makers. Still, the benefits of participatory processes, if it is assumed to be possible to avoid the problems, are multiple and justify experimentation. Citizens possess detailed data about the problems and potentials of areas under planning processes that no other analysis can reveal. In order to distil the benefits of the participative model from its problematic entirety, we must take into consideration the deficiencies presented by deliberative democrats. One possible way to overcome these problems is found in online communication. Many online platforms already provide venues for different participants (visitors, members, clients, donors etc.) and different forms of participatory activity (connecting with organizations, searching for information, learning about a policy field etc.) (Bruszt, Vedres and Stark, 2005). Without dealing with other problems or potentials of the participative model, the aim is to examine which forms of online communication can help to extract the above stated benefits from the participative model, while annulling the specific problems presented that the model has.

## Types of online communication

People use the Internet on a daily basis for a wide range of activities that could be considered to be "participation"—communicating, seeking information, content creation and so on (Livingstonea, Bobera and Helspera, 2005). The content is no longer created and published by individuals, but instead is "continuously modified by all users in a participatory and collaborative fashion" (Kaplan and Haenlein, 2010, 61). A vast amount of literature already covers different characteristics of online communication for participation in politics and other spheres of public life (Shah, Cho, Eveland and Kwak, 2005; Bakker and de Vreese, 2011; De Zúñiga, Puig-I-Abril and Rojas, 2009). Online communication tools that can be used for participation purposes (i.e., collecting information) have the characteristic of enabling the user to transfer information to others—in private or public settings. Also, this information must remain visible for a limited or unlimited time, sufficient for others to see it and/or note it. These are synchronous online communication tools (instant messaging, chats, conferences etc.), asynchronous (e-mail, thread discussions, blogs, wikis etc.) or hybrid forms (collaborations etc.). They demand different kinds of commitment and, due to their characteristics, provide different forms of information. We will compare ways of receiving data from citizens through new online technologies (which do not demand investment of time for travel from one place to another to participate) with problems of the participatory model.

**Users' will to participate in online communication:** People use online communication when other obligations allow them to and from the intimacy and comfort of their houses or workplaces, most of the time sharing and receiving information through some of the tools mentioned—posting comments, chatting, sharing documents, blogging and so on. All of these tools are accessible, fast, inexpensive and de-territorialized, with reduced formality and increased freedom and ease of use (Bruszt, Vedres and Stark, 2005; De Zúñiga, Puig-I-Abril and Rojas, 2009) in contrast to public discussions organized in public spaces and buildings. People use social networks on daily bases and comment on things they wouldn't bother to find and comment upon elsewhere (even on the Internet)—it is easy and accessible.

Since people are already using the Internet and searching through different data, they are likely to see and/or join more discussions than they would if they were invited to participate in a civic hall, for example. McKenna and Bargh (2000) stress anonymity as an attractive online communication feature, along with choice about when to participate and the control one has, in the form of taking time to think about on-going correspondence. Users have a choice as to which subjects they will comment on, which debates they wish to join, what they wish to learn and with whom they want to communicate. Different forms of communication tools enable them to choose how much they want to contribute: that is, how much time they wish to invest in relation to the information they want to transfer.

**Majority strength and the psychology of the masses in online communication:** The boundaries delineated by cultural constructs of race, gender, social position, authority, appearance and so on can be by-passed in online communication to create a clear interaction (Reid, 1991). In this way the system gives voice to each individual interested, instead of to just those who are the loudest in the crowd, and each argument receives an equal amount of attention (being a way of communicating where everyone has the same chance to speak or be heard). The chances of individuals leading a great number of people from the mass to support their views and influence others are thus smaller. Different types of information exchange support this benefit. When citizens do not wish to state their opinion publicly, they have the option of sending personal messages in online communication—asking questions or sharing data through emails and instant messaging in relative privacy. Kiesler, Siegel and McGuire (1984) have described a distinct feature of social anonymity of online communication in comparison to conventional forms of interaction. Since they are anonymous, users of online communication tools thus behave in a more uninhibited manner than they would in face-to-face communication—without social context cues to inhibit a free exchange between people (Reid, 1991).

**Self-development through online communication:** People already use different forms of online tools for learning or receiving information. Chats and threaded discussions enable fast information dissemination and exchange. Conferences enable the transfer of more complex knowledge in an attractive way, from conference organizer to public, with the possibility of commenting. Blogs enable each citizen to elaborate on the matter of their interest and share opinions on it through comments and sharing links to other blogs. They are useful for disseminating the latest information, creating an environment open for conversation about different topics, and for more intensive information flow and learning (Bransford, Brown and Cocking, 2000). Wikis can help gather and receive information on a given subject from a larger number of interested parties, gathering information in one place, and affording reliability checks from citizens themselves (Smith, Mills and Myers, 2009). The educative element of online communication is most obvious in the case of "lurks"—people who join a community and do not post, but search for information and learn (Nonnecke, Andrews and Preece, 2006, 7).

## Avoiding utopian presumptions about online communication

Despite this overheated rhetoric, it is also necessary to observe the down-sides of online communication tools if they are to be used in planning, in order not to fall into another utopian model. The most observed problems with participation are people's motivation to react, even from their homes, and finding ways of motivating them. People comment on and read about the subjects which are of direct interest to them (for amusement or information). This can be viewed as advantageous for planning, since only interested parties will react on the subject, but can be a draw-back, since less people means less information on the planning area and on decisions made. Unfortunately, the potential to create platforms with wide, active and loyal communities is not always achieved. Many communities fail and a variety of communities suffer from a deficit of visible content contribution.

Many authors deal with this problem using social science research and finding different modes of motivating people to interact. Social scientists point to varied and multiple factors influencing one's online contribution to a group, such as: the size of the group, the group and its members' attractiveness, expectations of performance, the importance of contribution outcomes, incentives, and the probability for interacting again (Karau and Williams, 1993; Oliver and Marwell, 1998). Ludford et al. suggest that community members "like receiving information about the unique perspective they bring to the group and participate more because of it" (2004, 7). Scandalous or controversial topics are effective tools for stimulating participation (Guerin, 2003), which is an advantage for planning where most situations are controversial (concerning politics, economy, design etc.). This can be an advantage, but many problems are attached to this kind of information sharing. According to Ludford et al. (2004, 7) "flame wars" should be avoided, since they bring more "heat than light", but can encourage discussion when disagreement is permissible. "Flaming", the expression of anger, insults and hatred, is a common phenomenon in all forms of computer-mediated communication, but social sanctions are present and operators have the ability to "kill" users (Reid, 1991). The role of operators is multiple and can be of utmost importance for motivating the community to participate, through their roles of conversation stimulator, conflict resolver, summarizer of debates, supporter, cleaner and so on (Davies & Chandler, 2011). Preece and Shneiderman (2009) have made their contribution to the subject of motivating users by gathering research on the subject of how to motivate different types of users (readers, contributors, collaborators or leaders of an online community) and what to do to move them from one level to another. Suggested solutions are numerous.

While the use of attractive, playful applications is most often successful, there are fewer stories about durable large-scale successes that deal with difficult issues such as crime reporting and disaster response (Preece and Shneiderman, 2009) or similar issues that use the same type of participation needed for planning. According to most authors, if civic-minded technology specialists devote sufficient attention to the problems of motivation, they can be reduced (by carefully choosing subjects, giving rewards and recognition, engaging operators etc.). The opportunity to produce startling changes in online approaches to planning participation means that "state and local agencies should begin their own pilot projects" (Preece and Shneiderman, 2009, 26), but small steps are feasible and can be explored.

## References

Bachrach, P. & Botwinick, A. 1992. *Power and empowerment: a radical theory of participatory democracy*. Philadelphia: Themple University Press.

Bakker, T. P. & de Vreese, C. H. 2011. "Good News for the Future? Young People, Internet Use, and Political Participation." *Communication Research* 38: 451-470.

Benhabib, S. 1996. "Toward a deliberative model of democratic legitimacy." In *Democracy and Difference: Con-testing the Boundaries of the Political* edited by Shaila Benhabib. Princeton: Princeton University Press.

Bohman, J. 1996. *Public deliberation: pluralism, complexity, and democracy*. Cambridge: M.I.T. Press.

Bransford, J. D., Brown, A., & Cocking, R. (Eds). 2000. *How People Learn: Mind, Brain, Experience and School*, Expanded Edition. Washington, DC: National Academy Press.

Bruszt, L., Vedres, B. & Stark, D. 2005. "Shaping the Web of Civic Participation: Civil Society Websites in Eastern Europe." *Journal of Public Policy* 25, no. 1: 149-163.

Cohen, J. 1997. "Procedure and substance in deliberative democracy." In *Deliberative democracy: Essays on reason and politics* edited by James Bohman and William Rehg. Cambridge: M.I.T. Press.

Davies, T. & Chandler, R. 2011. "Online Deliberation Design: Choices, Criteria, and Evidence." In *Democracy in Motion: Evaluating the Practice and Impact of Deliberative Civic Engagement* edited by Tina Nabatchi, Michael Weiksner, John Gastil, and Matt Leighninger. Oxford University

De Zúñiga, H. G., Puig-I-Abril, E. & Rojas, H. 2009. "Weblogs, traditional sources online and political participation: an assessment of how internet is changing political environment." *New Media Society* 11: 553-574.

Dye, T. & Zeigler, H. 2008. *The irony of democracy: an uncommon introduction to American politics*. Belmont: Wadsworth Publishing Company.

Fischer, F. & Forester, J. (Eds). 1993. *The argumentative turn in policy analysis and planning*. Durham, N.C.: Duke University Press

Guerin, B. 2003. "Language use as social strategy: A review and an analytic framework for the social sciences." *Review of General Psychology* 7, no. 3: 251-298.

Guthmann, A. & Thompson, D. 1996. *Democracy and disagreement*. Harvard College: President and Fellows.

Guthmann, A. 1993. "Democracy." In *A companion to contemporary political philosophy* edited by Robert Goodin & Philip Pettit. Oxford: Blackwell.

Hauptman, E. 2001. "Can less be more? Leftist deliberative democrats' critique of participatory democracy." *Polity* 33, no. 3: 397-421.

Healey, P. 1992. "Planning through debate: a communicative turn in planning theory." *Town planning review* 63, no. 2: 143-162.

Innes, J. E. 1995. "Planning theory's emerging paradigm: communicative action and interactive practice." *JPER* 14: 183-188.

Kaplan, A. M. & Haenlein, M. 2010. "Users of the world, unite! The challenges and opportunities of Social Media." *Business Horizons* 53: 59-68.

Karau, S.J., & Williams, K.D. 1993. "Social loafing: A meta-analytic review and theoretical integration." *Journal of Personality & Social Psychology* 65, no. 4: 681-706.

Kiesler, S., Siegel, J. & Mcguire, T. W. 1984. "Social Psychological Aspects of Computer-mediated Communication." *American Psychologist* 39, no. 10

Livingstonea, S., Bobera, M. & Helspera, E. J. 2005. "Active participation or just more information? Young people's take-up of opportunities to act and interact on the Internet. Information," *Communication & Society* 8, no. 3

Ludford, P. J., Cosley, D., Frankowski, D. & Terveen, L. 2004. "Think Different: Increasing Online Community Participation Using Uniqueness and Group Dissimilarity." Proceedings of the *SIGCHI Conference on Human Factors in Computing Systems*, ACM New York, NY, USA pp. 631-638

Mansbridge, J. 1999. "Altruistic truth." In *Democracy and trust* edited by Mark Warren. Cambridge: University Press.

McKenna, K. Y., & Bargh, J. A. 2000. "Plan 9 from cyberspace: the implications of the Internet for personality and social psychology." *Personality and Social Psychology Review* 4: 57-75.

Nonnecke, B., Andrews, D. & Preece, J. 2006. "Non-public and public online community participation: Needs, attitudes and behaviour." *Electron Commerce Res* 6: 7-20.

Oliver, P.E., & Marwell, G. 1998. "The Paradox of Group Size in Collective Action: A Theory of the Critical Mass. II." *American Sociological Review* 53, no. 1: 1-8.

Preece, J. and Shneiderman, B. 2009. "The Reader-to-Leader Framework: Motivating Technology-Mediated Social Participation." *AIS Transactions on Human-Computer Interaction* 1, no. 1: 13-32

Reid, E. 1991. "Electropolis: Communication and Community on Internet Relay Chat." University of Melbourne, Department of History http://www.irchelp.org/irchelp/communication-research/academic/academic-reid-e-electropolis-1991.html (02.02.2014)

Shah, D. V., Cho, J., Eveland, W. P. & Kwak, N. 2005. "Information and Expression in a Digital Age: Modelling Internet Effects on Civic Participation." *Communication Research* 32: 531-565.

Smith, E. J., Mills, J. E., & Myers, B. 2009. "Using wikis and blogs for assessment in first-year engineering." *Campus-Wide Information Systems* 26, no. 5: 424-432.

Warren, M. 1992. "Democratic theory and self-transformation." *The American political science review* 86, no. 1: 8-23.

# CHAPTER THREE

# PLACE COMPETITIVENESS EXPRESSED THROUGH DIGITAL DATA

MILENA VUKMIROVIC[1]
AND EVA VANISTA LAZAREVIC[2]

**Abstract** | The globalized and decentralized world that we live in is simultaneously an arena for the competition of cities belonging to a global network. Striving to secure a better position on the international scene, responsible managements turn to entrepreneurial activities aimed at finding new forms of competitive capitalism. All efforts are directed towards improving place identity with the aim of building a strong, positive image, and therefore representing a way in which a certain place is experienced in one's consciousness. Based on the place image model of communication, which understands primary, secondary and tertiary communication, this paper aims to give an overview of contemporary techniques and tools to be used for tracking tertiary communication, as a key element of the process of building strong place image. Tertiary communication is followed indirectly, not by analyzing the contents of a text, a photo, or a video published online, but rather by analyzing geospatial the metadata that is shown on maps, and also connected to other forms of published content. The presented maps are formed by taking over geospatial metadata, i.e. the geo-tags from five relevant internet platforms: Panoramio Google Service, Instagram, Flickr, Twitter and Foursquare.

**Keywords** | Place image, Tertiary communication, Photo sharing, Micro-blogs, Geo-tagging,

---

[1] Research Associate, University of Belgrade – Faculty of Architecture, Belgrade, Serbia, milena.vukmirovic@urbanlab.org.rs
[2] Professor, University of Belgrade – Faculty of Architecture, Belgrade, Serbia, eva.vanistalazarevic@gmail.com

## Introduction

The model for identity-communication is characterized by a pragmatic approach. Such pragmatism is reflected in a clear display of the inseparability of identity, communications and image in providing competitive advantages, and in describing the key role of three components of a municipal communication system. It is based on the model which develops connection between corporate identity and the process of corporate communications (Balmer and Gray 2000). The basic advantage over the stated models is reflected by showing communication as a three-fold bridge between the identity of a city and its resulting image and reputation. The second feature of the model is reflected in putting emphasis on the general population. This does not mean that the other target groups are less important, but rather, merely stresses the fact that all activities are performed in the name of inhabitants of a given city, and that the final goal is an overall improvement to their quality of life.

This model points to the specific importance of identity and communication, but also to the change in the way that these elements are treated, from functional to strategic. Observing city communication as a three-fold system encompasses *primary, secondary and tertiary communication*. Primary communication is defined through presenting a positive image and framework for creating a strong reputation. Secondary communication is further shaped in order to support and strengthen this primary communication. Finally, tertiary communication is expected to both be positive and to achieve a result in terms of a premium reputation, having in mind that the processes of both primary and secondary communication have been already adequately conceived and realized by the point at which tertiary communication would be considered. Kavaratzis (2008) argues that all encounters and confrontations with a city are realized by perceptions and image. In accordance with this, it is claimed that it is also necessary to plan the image of a city itself, as well as to (Vermaulen 2002) market it.

The Primary form encompasses the communication effects of city activities, those through which the communication itself is not the aim, but a byproduct. It is divided into four broader areas of intervention: landscape strategies, infrastructure projects, organizational and administrative structure, and the city's behavior. Secondary communication represents formal, intentional communication which is found in well-known marketing instruments, such as in-door and out-door branding, public relations, graphic design, the use of logos, etc.

Tertiary communication is based on **word of mouth**, and is strengthened by the use of media and through communication by competitors. It is connected to the previous two modes of communication by an interrupted line however, and it cannot be influenced by those participating in the marketing process. *The entire process of branding and other two types of communicating image aim to encourage and strengthen positive tertiary communication* (Kavaratzis 2008, 41-43). This is especially applicable to the social circles of the population, which are at the same time considered to be the most important target audience in city branding, as well as being the most important participants in place marketing.

If traditional diplomacy is characterized by government-to-government (G2G) relations, and public diplomacy as being government to people (G2P), then through having in mind the importance of live words, place branding should include a 'people to people' (P2P) relationship. By establishing this relationship, we achieve a result in which *all citizens become organized in such way to be a voice of quality and value of a city, generating such medium of advertising which is equal to an enormous task of communicating something so complex, yet directed towards large number of people* (Anholt 2007, 105).

Tertiary communication appears in another form, one which gains more importance every day and threatens to jeopardize the dominant influence of mass media. It is a form marked by the term 'word of mouth', and is achieved through the kind of communication that is enabled by social networks, blogs and micro-blogs (Facebook, Twitter, LinkedIn, etc.). These platforms function on the principle wherein an individual will open a personal account, and thus enable said individual to connect to people and organizations based on mutual choice. In such a way, a person selectively picks information in accordance with personal affinities in relation to other persons and institutions.

In this paper, the emphasis is put on geospatial data which can be connected to content on social networks or portals through the exchange of photos. Presenting the retrieved geo-data on a map has opened a new dimension in analyzing the ways in which a city is used, based on real spatio-temporal data. By careful observation and the interpretation of the results of these analyses, it is possible to form 1) a ranking list of the most visited cities and their specific locations, 2) the separate locations within a city which are most visited by tourists, i.e. inhabitants, 3) the ways in which people move throughout the city, 4) separate the most visited locations and the periods during the day when that visit occurs, 5) map the attractiveness of the specific attributes of certain locations and the temporal interval in which these attributes are most visited, etc.

## Following place attractiveness through geo-tags

The general definition of attractiveness states that it is a quality of a person or an object which generates interest from other people. By observing it in relation to a place, attractiveness quantifies how much something is able to attract the attention, and influence the decisions, of one or more individuals, (Uchino et al. 2005) and can help to explain a variety of spatial-temporal phenomena (Furtado, Fuleto and Chiara, 2013). Various methods were used to express attractiveness in a quantitative way[3], while the very phenomenon itself was the subject of action for various disciplines: urban design, traffic, marketing, sociology, tourism, etc.

In the domain of urban design, Jan Gehl has equated the measure of attractiveness of certain spaces to the overall quality of a location (public open space). By establishing a correlation between the quality of a particular public open space and the number of activities that people carry out in them, Gehl noticed that by increasing the quality of outer environment the number of follow-up activities increased as well (Gehl 2008). The stated correlation is presented in a diagram (Fig. 1) where it is seen that, independently from the quality of the outer environment, the level of ‘necessary’ activities does not change, but by increasing this quality, the level of ‘follow-up’ activities increases dramatically, which causes an overall increase in social activities. Social activities[4] are the fruit of quality and the duration of the other two types of activities, because they happen spontaneously during encounters.

The development of technology and the increasing number of smartphone users (and users of other electronic mobile devices) has enabled new ways of measuring the attractiveness of cities and specific locations, based on the intensity of visits. As a key resource, one can single out the spatio-temporal metadata – the geotag – which is linked to devices themselves, but also to the content created and published on the internet using various social networks.

[3] Primarily the Gravitational Attractiveness Model by Reilly and the Theory of Central Places by Christaller.

[4] They do not include children playing, greetings and conversation, mutual activities of various kinds or simply observing and listening to other people.

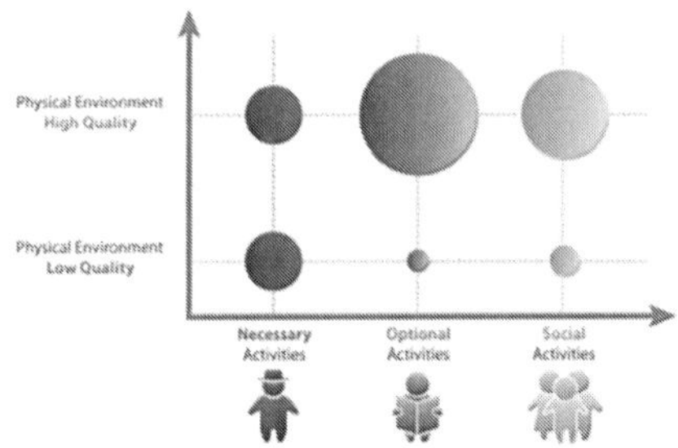


*Fig. 1: Diagram – place quality vs. number of activities. Source: Gehl 2008*

## Geotagging

Geotagging is the process of adding geographical identification metadata to various media, such as a geotagged photograph or video, websites, SMS messages, QR Codes or RSS feeds, and is a form of geospatial metadata. This data usually consists of latitude and longitude coordinates, though they can also include altitude, bearing, distance, accuracy data, and place names. It can help users find a wide variety of location-specific information. For instance, one can find images taken near a given location by entering latitude and longitude coordinates into a suitable image search engine. Geotagging-enabled information services can also potentially be used to find location-based news, websites, or other resources. Geotagging can tell users the location of the content of a given picture or other media, or the point of view, and conversely, on some media platforms, can show media relevant to a given location.

Images are one of the oldest types of media to use geotagging. Certain formats (like the JPEG format) allow for geographical information to be embedded within the image and then read by picture viewers, which allows the exact location of where a picture was taken to be saved with a photograph. Smartphones with built-in cameras and GPS have made geotagging images a breeze. Some blogging and microblogging sites such as Twitter also support geotagging. This is especially popular with blogging and social networking sites that cater to a mostly mobile audience.

Having in mind the stated possibilities, i.e. the sources of spatio-temporal metadata, 8 case studies are selected in order to shed some light on the potential of this information, and the possibility of its interpretation. Case studies are grouped according to the internet platform from which the geospatial data is used.

## Google Maps Panoramio Service

Panoramio is a geolocation-oriented photo sharing website. Accepted photos that are uploaded to the site can be accessed as a 'layer' in Google Earth and Google Maps, with new photos being **added** at the end of every month. The site's goal is to allow Google Earth users to learn more about a given area by viewing the photos that other users have taken at that place. Based on the data taken from the panoramio web site, a Sightmap was formed in order to better represent the attractiveness of certain locations, based on the number of photographs taken at that location. The heatmap shows the places people "like" based on the number of Panoramio photos at each place in the world. The dark areas have few photos, the red areas have more and the yellow areas have a large number of photos geotagged.

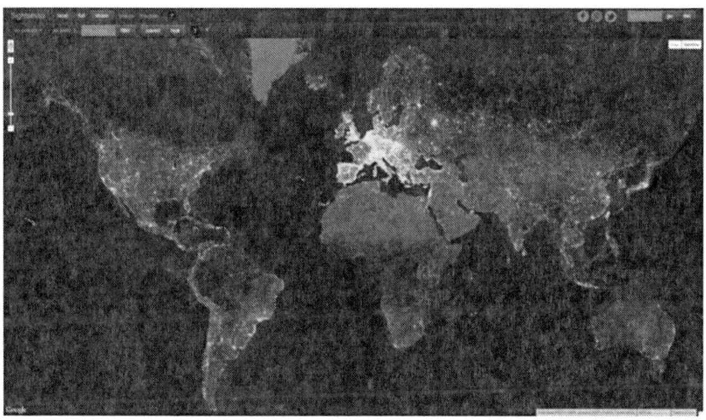

*Fig. 2: Panoramio photos world heatmap. Source: http://www.sightsmap.com/#*

Maps are available for top 15000 places in the world, sorted by the number of photos in an area of a size of a few square kilometers around the place center. The popularity ranking of places in high-res area maps is computed by combining place hotness with popularity rankings from Wikipedia, Foursquare and real-time Google places selection.

The lists of popular places were formed based on the data garnered in this way, wherein the popularity was measured based on the number of photographs taken whose geographical data is linked to the location in question. According to these data, the first five most attractive cities are: 1. New York, 2. Rome, 3. Barcelona, 4. Paris and 5. Istanbul. All of the stated (and other) cities can be analyzed in more detail by determining the specific locations in the city itself that are most attractive. By analyzing New York (Fig. 3), as the most popular city according to this study, it was determined that its attractive locations are: 1. Solomon R. Guggenheim Museum, 2. Tweed Courthouse, 3. Empire State Building, 4. Battery Park and 5. Grand Central Terminal Park Avenue Viaduct.

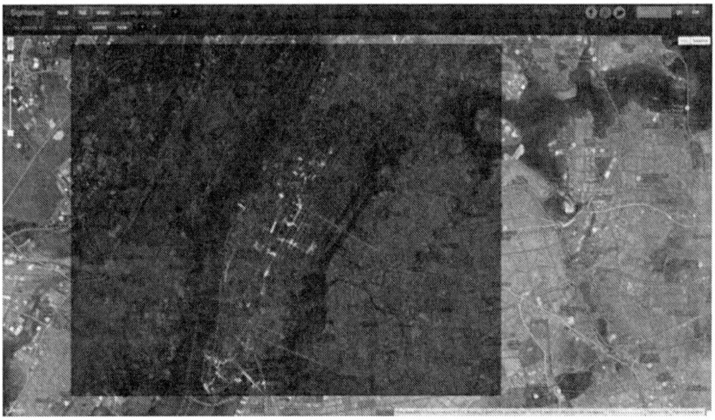

*Fig. 3: Panoramio photos New York heatmap. Source: http://www.sightsmap.com/#*

Further research enables the analysis of the content of those photos, which can be used to interpret ways in which users and visitors experience this space, what attracts them and what they recommend to other users.

## Instagram

Instagram is an online photo-sharing, video-sharing and social networking service that enables its users to take pictures and videos, apply digital filters to them, and share them on a variety of social networking services, such as Facebook, Twitter, Tumblr and Flickr. This photo-sharing app showed strong growth in 2013. While Facebook showed 3-percent growth in active users, the behemoth social network's photo-sharing sister service gained 23 percent in active users (Knibbs, 2014).

By analyzing the content which appears most on uploaded photographs, it is observed that the largest number of users of this social network take photographs of food or pets. However, the third most generated content is cities – interesting happenings, places, installations etc. Keeping this in mind, Instagram has formed a list of the most popular locations and cities. Based on the trend of appearance of certain locations, a list of the most-instagrammed cities was formed. The company has up until now published lists for 2012 and 2013. According to the data published in January 2014, the five most popular cities are 1. New York, 2. Bangkok, 3. Los Angeles,

4. London and 5. Sao Paolo (Instagram 2013). Among the five most popular city locations in 2013 are: 1. Siam Paragon Shopping Mall u Bangkok, 2. Times Square in New York, 3. Disneyland in California, 4. Bellagio Fountains in Las Vegas and 5. Disney World in Florida (Instagram 2013).

*Fig. 4: Instagram photos taken on Times Square. Source: http://www.gramfeed.com/instagram/places/3001373*

Besides providing an overview interpreted in the form of a list, the analysis can be done by memorizing the very content that is published on the photographs, which is also linked to a specific location (Fig. 4).

## Flickr

Flickr is a photo-sharing web platform that allows its users to store, sort, search, and share photos online. For this research, the sets of photos taken in 2010 by Eric Fischer carry a huge weight. The taken spatio-temporal metadata[5] from photos published on Flickr (and Picasso, also a photo-sharing web platform), Fischer uses and creates several hundred maps of cities which he then divides into three groups, depending on the manner of interpretation selected for the results gathered: *Geotaggers World Atlas*, *Locals and Tourists* and *See something or say something*. Within these thematic sets, the maps are ordered by the number of pictures taken in the central cluster of each one – which has at the same time enabled the ranking of selected cities. According to the opinion of the author, in this way cities with polycentric character are somewhat neglected if this methodology is applied (Tokyo and Los Angeles) meaning that they probably get a lower placement than they really deserve because there are gaps where no one took any pictures. The cities which are not on the list did not possess enough adequate data for a good quality map to be created. All the maps are to the same scale (a square measuring 15 miles on each side), which was chosen to be just large enough for the central New York cluster to fit.

[5] The photo locations come from the public Flickr and Picasa search APIs.

The sets entitled *Geotaggers' World Atlas* and *Locals and Tourists*[6] (Fig. 5) created during 2010 have attracted much attention from professionals, and have from the very start been understood as maps of human interest, depicting a city's most photogenic and commonly visited sights. The color was used to separate spatial data in relation to the temporal interval in play when the photos were made. Such analysis enabled us to determine which mode of transportation each photographer presumably used, considering the time stamps and distance travelled between a user's pictures. The information is color coded on each map (Fig. 5a): black is walking (less than 7mph), red is bicycling or a mode of transportation of equivalent speed (less than 19mph), blue is motor vehicles on normal roads (less than 43mph); green is freeways or rapid transit.

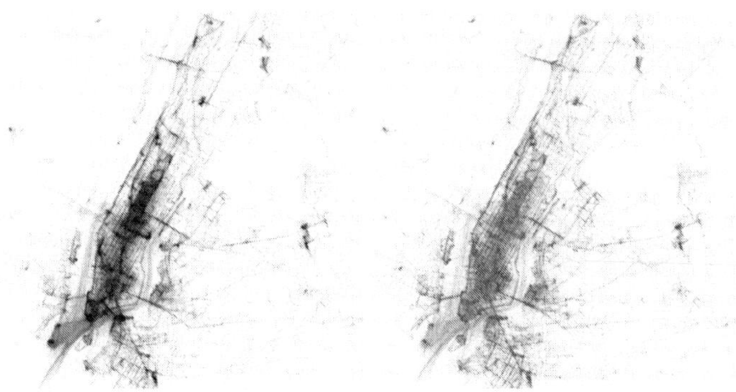

*Fig. 5: Eric Fischer's maps of New York: a) Geotaggers' World Atlas and b) Locals and Tourists. Source: Eric Fischer's photo stream on Flickr*

The second set of maps were the maps of tourism. Some cities like Las Vegas and Venice do seem to be photographed almost entirely by tourists. Others seem to have many pictures taken in places that tourists do not visit. Photographs taken over the course of a month or more are designated by blue plots on the map and represent locals. Photographs taken over the course of a month or less were presumably taken by tourists and are represented by red plots on the map. Yellow plots are designated for the unknown.

[6] *By using the same methodology, but using data collected from Twitter and in cooperation with Gnip team, Locals & Tourists maps were created, published in the Mapbox platform.*

## The geography and the topography of tweets

Twitter is an online social networking website and microblogging service that allows users to post and read text-based messages of up to 140 characters, known as “tweets”. *With its open API, Twitter has become one of the most popular data sources for social research, yet the majority of the literature has focused on it as a text or network graph source, with only limited efforts to date focusing exclusively on the geography of Twitter, assessing the various sources of geographic information on the service and their accuracy* (Leetaru and al. 2013). The authors report that few studies make use of the geographic information attached to tweets, while papers like Poblete, *et al.* (2011) have used it primarily as a filtering mechanism rather than focusing on the geography itself. Geotagging occurs when Twitter users make their location available so that others can see exactly where the Tweet was sent from. Only a small proportion of all Tweets are geotagged, but they provide *a “gold standard” dataset, helping us gain even deeper insights into the way people Tweet* (Leetaru and al. 2013).

The researches made by Miguel Rios (2013) from the Twitter Visual Insights Team are particularly significant for researching the attractiveness of cities and their locations. By visualizing the geotagged tweets (Fig. 6) published from 2009 to 2013, the contours of “active” counties, cities, locations, main roads and even waterways (fluvial and maritime traffic) have been outlined.

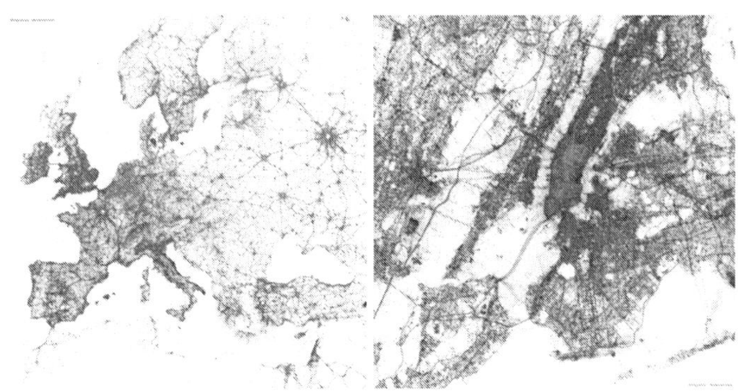

*Fig. 6: Map of the tweets since 2009: a) Europe and b) New York. Source: Rios 2013*

The second important research carried out at the same time is titled “Using Topography of Tweets”. In difference to “geography of tweets”, which portrayed the concentration using various shades on certain locations, these diagrams do so by introducing the third dimension (Rogers 2013). In such a way, a new topography of locations was developed, reflecting not the natural one, but rather the Twitter landscape. The peaks represent the places most Tweets are sent from, the troughs the fewest. Explore New York closely and you can pick out the Brooklyn and the Queensbury bridge — even the Staten Island ferry (Fig. 7).

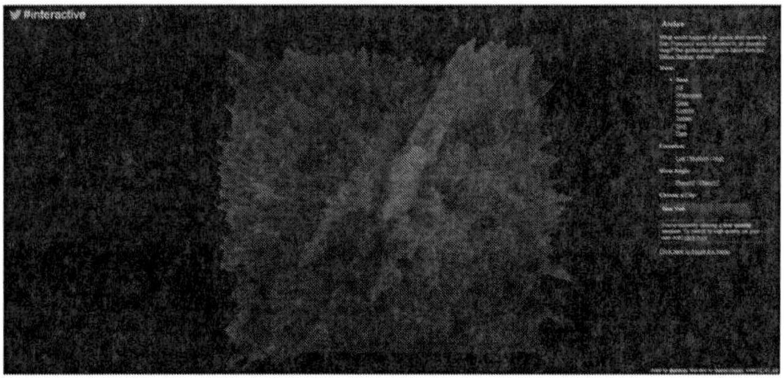

*Fig. 7: New York's topography of tweets. Source: Rogers 2013*

Fischer used red dots for Flickr pictures, blue for Twitter and white for both – all of this was publically available data in the Flickr and Twitter APIs. To some extent, the density reflects the density of footfall in different parts of the city, but with Flickr, geographic visualizations tend to reflect a more cultural consensus of photogenic places or tourist spots.

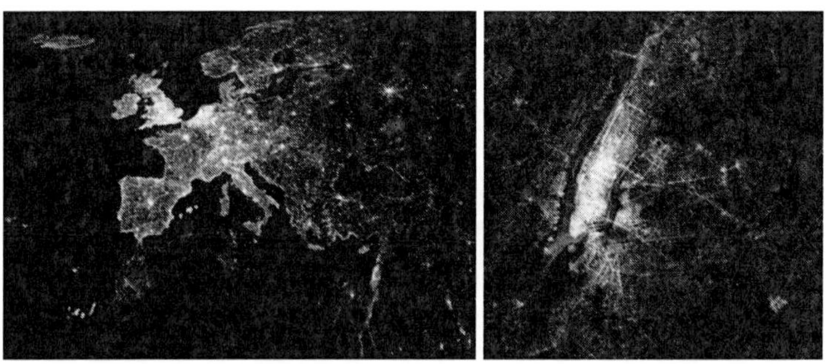

*Fig. 8: See something or say something in a) Europe and b) New York. Source: Eric Fischer's photo stream on Flickr*

The research has demonstrated that the geographies do not overlap because the places people take pictures of are not the places they tweet from. That is, people take pictures of tourist destinations; people tweet from where they live. It is no less than a visual map of how differently cities look when you view them through the eyes of a tourist or a native.

## Foursquare

Foursquare is a location-based social networking website for mobile devices, such as smart phones. Users "check in" at venues using a mobile website, text messaging or a device-specific application, by selecting from a list of venues that the application locates nearby. The location is based on GPS hardware in the mobile device or on a network location provided by the application, and the map is based on data from the OpenStreetMap project.

Having in mind the specificity of this social network, and the analytical possibilities which it offers, a selected research followed the 24 hour activity in six cities (New York, Tokyo, London, Istanbul, Chicago and San Francisco) in relation to the character of a location where users checked in. By observing the diagram, the chart shown on the lefthand side, one can follow which activity is most present, and which location is most visited, in each period during the day. It gives a better sense of why Foursquare's trove of data is such a treasure: *It not only indicates where people are but, more significantly, how, when, and why they're going there* (Foursquare 2013).

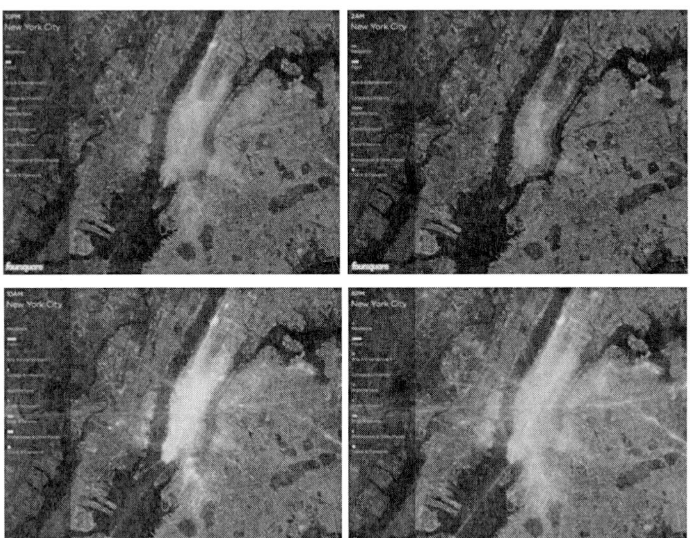

*Fig. 9: New York activity at a) 10PM, b) 2AM, c) 10AM and d) 6PM. Source: Foursquare 2013*

Color-coded signals show at what times of day, and by what means, users travel to various locations. In New York, for example, you can see yellow streaks of light zoom into Manhattan as users commute to work, and then slip away as they head home; at night, the city lights up in blue as more users head to nightlife spots. The researchers used the methods of comparative analysis, where they compared the ways in which citizens of New York and Tokyo, for example, use their cities, following the same temporal interval throughout the day.

## Conclusions

The demonstrated case studies are only a segment of the published research on an enormous amount of data, which can be obtained and interpreted by downloading it from various social networks. This paper solely selected the research based on geotagging, and that presented by using maps, as a means of presenting spatio-temporal data.

Observed from the aspect of urban design, the contemporary period, also characterized as the period of iPhone revolution, has created new perspectives in the real-time analysis of the different ways in which cities function. Each of the five selected platforms, which have the character of a social network, can be observed as means of tertiary communications, especially in two ways. The first way comes down solely to the position data which the geotagged information provides, whose number can be, in the widest sense, understood as the high popularity of a certain location being directly proportional to its attractiveness. The second way takes into account the temporal dimension of the concrete spatial record, which is later interpreted in the form of the speed of movement and interval of presence at each location. When the information on the character of certain locations is added to that, the way of life of a city, and the preferences of its inhabitants, can be easily analyzed.

In the context of place attractiveness which is quantified in this way, the general interpretation can be that a geotag confirms someone's presence at a certain location in a certain time, which is also a confirmation that at that place, at that time, something interesting happened. *However, when that is recorded with a photo, this can be interpreted as if it gave birth to a wish to document it* (Fisher 2010).

Research conducted on the basis of these data may indicate the success of certain interventions in space from the aspect of urban design; it can determine the value of a particular location for economical considerations, or the ways people use the city from the point of view of urban sociology.

Depending on the availability of these enormous amounts of data taken from social networks, and the creativity of their presentation and interpretation, a wide field of studies on the city and the ways in which it functions become possible.

## Acknowledgements

The paper was realized as a part of the research project entitled “Modernization of Western Balkans”, No. 177009 and the project “Spatial, Environmental, Energy and Social Aspects of Developing Settlements and Climate Change – Mutual Impacts”, No TP36035, PP1: “Climate change as a factor of spatial development of settlements, natural areas and landscapes”, both financed by the Ministry of Education, Science and Technological Development of the Republic of Serbia.

Paper is presented on the 1st International Academic Conference on Places and Technologies held at 3rd and 4th April 2014 in Belgrade, Serbia.

## References

Anholt, S. 2007. *Competitive identity: A New Brand Management for Nations, Cities and Regions*. Hampshire/New York: Palgrave MacMillan

Balmer J., Gray, E. 2000 "Corporate identity and corporate communications: creating a competitive advantage", *Industrial and Commercial Training*, Vol. 32 No. 7, pp.256 – 262

Fischer E., 2011. “Language communities of Twitter,” at http://www.flickr.com/photos/walkingsf/6277163176/,

Fisher, E. 2010. “Locals & Tourists”. Accessed February 10, 2014. https://www.mapbox.com/labs/twitter-gnip/locals/#9/41.1683/28.9902,

Foursquare. 2013 “Foursquare check-ins show the pulse of cities”. Accessed January 15, 2014. https://foursquare.com/infographics/pulse

Furtado, A. S., Fileto, R. and Renso C. 2013 “Assessing the Attractiveness of Places with Movement Data” *Journal of Information and Data Management*, Vol. 4, No. 2 (June): 124–133.

Gehl J. 2008. *Life Between Buildings: Using public space*. Copenhagen: Danish Architectural Press,

Instragram, 2013 “Top locations on Instragram in 2013”, *Instragram news*, Accessed February 10, 2014, http://blog.instagram.com/post/69877035043/top-locations-2013

Karavatzis, M. 2008. "From City Marketing to City Branding: An Interdisciplinary Analysis with Reference to Amsterdam, Budapest and Athens." *Doctoral dissertation*. Groningen: University of Groningen,

Knibbs, K. 2014 "Instagram is growing faster than Twitter, Facebook, and Pinterest combined", *Digital Trends*, January 21. Accessed February 10, 2014, http://www.digitaltrends.com/social-media/instagram-is-growing-faster-than-twitter-facebook-and-pinterest-combined-in-2013/#ixzz2svybcZKR

Leetaru, K., Wang, S., Cao, G., Padmanabhan, A., & Shook, E. 2013. "Mapping the global Twitter heartbeat: The geography of Twitter". *First Monday*, 18 (5). doi:10.5210/fm.v18i5.4366

Poblete B., Garcia R., Mendoza M., and Jaimes A., 2011. "Do all birds tweet the same? Characterizing Twitter around the world," *CIKM 2011: 20th ACM Conference on Information and Knowledge Management*. Accessed January 16, 2014. http://www.ruthygarcia.com/papers/cikm2011.pdf

Ríos M. 2013 "The geography of tweets". Accessed January 15, 2014. https://blog.twitter.com/2013/the-geography-of-tweets,

Rogers S. 2013. "The topography of Tweets". Accessed January 15, 2014. https://blog.twitter.com/2013/the-topography-of-tweets

Uchino, A., Furihata, T., Tanaka, N., and Takahashi, Y. 2005. "Some Contribution toward Spatial Urban Dynamics". *Proceedings of the System Dynamics Conference*. Boston, pp. 1–7,

Vermeulen, M. 2002 'The Netherlands, holiday country', in Hauben, T., Vermeulen, M. and Patteeuw, V. (eds) *City Branding: Image Building and Building Images*, NAI Uitgevers, Rotterdam, The Netherlands.

# CHAPTER FOUR

# TRANSPARENCY OF SCALE: GEOGRAPHICAL INFORMATION PROGRAM (GOOGLE EARTH) AND THE VIEW FROM BEYOND

PAVLE STAMENOVIC,[1] DUNJA PREDIC AND DAVOR ERES

**Abstract** | 'Powers of Ten', a nine minute film directed by architects Charles and Ray Eames, visualized for the first time the journey through the various scales of existence – from outer space to the space of the atom. The film inspired the Keyhole Project, which twenty years later became Google Earth, an online tool for surveying the surface of Earth, making those intriguing images from the 1970s part of our everyday life.

Humanity has gained a new viewpoint, an overview of the territory of Earth. This paper argues that this transformation of the nature of the vista gives rise to the emergence of new modes of perception, cognition and experience of space and place specific to the emerging era (thus provoking new modes of production of space). This paper raises questions regarding the case of the Google Earth program and its impact on spatial perception in the context of architectural discourse.

**Keywords** | perception, scale, territory, perspective, photography, virtual

## Introduction: Perception of Space in the Digital Age

The paper argues for the significance of the emergence of Virtual Globe platforms such as Google Earth, and their effect on spatial practice

[1] PhD candidate, teaching assistant, Faculty of Architecture, University of Belgrade, Serbia, pavle.stamenovic@gmail.com

regarding perception and the causal nature of the relationship between technology and society. The significance of Google Earth goes beyond the mere utility of the software, as it is an important platform for determining contemporary perception. The new overview of Earth, combined with complex layers of data, follows previous shifts in perception: the appearance of central perspective in the Renaissance and the rise of photography in modern times. In this regard, the paper argues that it is the present mode of representation of space which has caused the latest shift in perception.

Until 20 years ago, architecture still possessed the apparatus and tools for conceiving and perceiving particular spatial ideas. But even more, the discipline of architecture held a sovereign position in projecting and imposing our spatial perception of the world. In today's digital age, characterized by user-friendly platforms for spatial manipulation, a shift in perception is inevitable; a shift from a world dominated by the architect as the master-designer of the image of space to an age in which micro-generated spaces, guided and maneuvered by consumers of open-ended platforms, prevail. These platforms promote interaction as a means of communication between the actual and the virtual, with the interface as the main surface of this interaction. The architectural principle of projection (conducted through drawings, plans and models) generating unitary perception is confronted by fragmentary individual interpretations of space.

## Shifts in Perception: Multiplying Reality

In order to fully comprehend Google Earth and transcend the simplified notion of a new outlook on reality, it is necessary to return to previous shifts in perception. In Renaissance times, Brunelleschi's 'invention' of linear perspective was more than an improvement in the technique of representation; for his contemporaries it was a manifestation and proof of the prosperity and inherent development of an enlightened and prosperous society. A new model for representing reality accurately was produced in order to more convincingly assure the viewer of the credibility of an architectural idea. Brunelleschi's 'invention' of perspective was the result of a constant quest for ever more convincing modes of three-dimensional visualization. Linear perspective on a 2D surface seduces the viewer into believing that a certain loss of 3D has been compensated for. A wish for seduction has led us to insist on the accuracy of representation.[2] However, this apparatus of 'seduction' produces 'side effects' which generate human sensitivity to the distinction between real space and realistic two-dimensional space.

With the invention of linear perspective as a way of seducing the observer, an image offers the additional value of depth to the surface. As Neil Leach observes in Baudrillard's writings, this seduction starts to vanish with the advances of the Industrial Revolution, as the contemporary world starts to appear: 'Seduction [perspective] removes something from the order of the visible, while production [photograph] constructs everything in full view, be it an object, a number or a concept.' Instead of a concept of seduction, there was progressive 'emphasis on production with its concern to make everything clear and objectifiable'.[3] A new era and a new shift in perception were therefore dominated by the invention of the camera.

The first photograph was a photograph of architecture.[4] Although the intention to capture a moment probably came from the idea of photographing a person, architecture and photography continued to develop together in an alternating, causal sequence. At that time, two types of architectural photography were typical, both primarily of documentary character: elevation as a representation of a drawing (i.e. a facade) and perspective as a representation of volume.

Photography had an enormous effect on the development of modern architecture. Architecture and photography worked together to create the identities of contemporary man, the contemporary building and the contemporary city. This was no longer merely for the sake of documentation. Together they caused a shift in the way architecture was seen (and therefore understood, promoted and advertised), and thus wanted. Images of architecture are used as statements and manifestos. They are used during the design process, as tools for conceptualisation, and in education, for the evaluation of work. Beatriz Colomina (1994)

[2] Baudrillard sees this as the logic of seduction, which he defines as 'that which extracts meaning from discourse and detracts it from its truth'. The real only remains true if we believe in fictions without functions. See: Mark Poster (ed.), *Jean Baudrillard, Selected Writings* (Stanford: Stanford University Press, 2001), p. 152.

[3] See Neil Leach, *The Anaesthetics of Architecture* (Cambridge, MA: MIT Press, 2000), p. 73.

[4] *View from the Window at Le Gras* by Nicéphore Niépche from 1826–27, taken from his window in a village in Burgundy, France, is the oldest surviving photograph.

claims that photography, rather then the real location of buildings, was the genuine site of Modernism.[5] Photography is responsible for the disappearance of a localised approach to design and the emergence of the International Style in architecture. Later, this correlation between architecture and image brought us to the Bilbao Effect and secured the new role of urbanism in globalization.

Alluding to Jules Marey's photographic gun, Bruno Latour suggests that the biggest flaw of buildings (and photography for that matter) is their static nature:[6]

> *Almost four centuries after perspective drawings and more than two centuries after the invention of projective geometry there is still no convincing way to draw the controversial space that a building almost always is. It is hard to believe that the powerful visualizing tools we now possess are still unable to do more than Leonardo, Duerer, or Piero. [...] We should finally be able to picture a building as a moving modulator regulating different intensities of engagement [...]*[7]

Inside this phenomenon, architects were compelled to adapt to the techniques of a film director in relation to space, life scenarios and the construction of spatial possibilities.

Modernists assigned great importance to the aerial view. Their relationship with zenithal perspective was described by Nadar,[8] creator of the very first aerial photograph, who said that from above cities and landscapes look like a toy-land, inviting one to play.[9] Modernists attached a negative connotation to the aerial view, as it depicted urban disorder, but at the same time they saw the aerial view as the perfect tool for arranging and controlling this chaos.

---

[5] For further reference see: Beatriz Colomina, *Privacy and Publicity: Modern Architecture as Mass Media* (Cambridge, MA: MIT Press, 1994)

[6] Latour, B and Yaneva, A, "Give me a Gun and I will Make All Buildings Move: An ANT's View of Architecture" in R. Geiser (ed.), *Explorations in Architecture: Teaching, Design, Research* (Basel: Birkhäuser, 2008)

[7] Ibid., pp. 80–89.

[8] A French journalist, photographer and balloonist who, due to his passion for new and expanded viewpoints, took the first aerial photographs in 1858.

[9] See Tanis Hinchcliffe "'The Synoptic View': Aerial Photographs and Twentieth-Century Planning" in Higgott, A and Wray, T (eds) *Camera Constructs: Photography, Architecture and the Modern City* (Farnham: Ashgate, 2012), p. 136.

As a result of technological developments resulting from both World Wars, the first photograph from space was taken in 1946.[10] Images taken from a missile showed 'how our Earth would look to visitors from another planet coming in on a space ship'.[11] We thus finally gained the perspective we desired, a *God's eye view*. One particular photograph changed the way we understand our world. A photograph named *Earthrise*, taken by William Anders during the Apollo 8 mission in 1968, is considered the most influential environmental photograph ever taken. Although the intention of the mission was to see and understand the Moon, the real discovery was our own home – Earth. As Robert Poole wrote in his book *Earthrise: How Man First Saw the Earth*, 'The space program, which was meant to show mankind that its home was only its cradle, ended up showing that its cradle was its only home.' It can be argued that this event was a turning point for the growing environmental awareness during the second half of the 20th century and until today, as well as resulting in the emergence of an environmental discourse in architecture.

*Fig 1. Earthrise - photograph by William Anders, NASA, 1968.*

---

[10] The photograph was taken using the V-2 missile, developed by German engineers during WWII for destructive purposes but later appropriated by Allied Forces.

[11] Clyde Holliday, the engineer who developed the camera, gave this description in *National Geographic* in 1950. (Source: http://www.airspacemag.com/space/the-first-photo-from-space-13721411/ Accessed: 02.02.2014)

Also in 1968, Buckminster Fuller's book *Operating Manual for Spaceship Earth* was published. 'We are all astronauts' he writes, on Spaceship Earth. In order to further convince us of this astounding idea, he continues with an explanation of the relativity of scale: 'Our little Spaceship Earth is only eight thousand miles in diameter, which is almost a negligible dimension in the great vastness of space [...]'.[12] It was as if scale-fever had seized everyone in 1968. The movie *Powers of Ten*, by architects Charles and Ray Eames, had created a revolution in scale perception. The journey from the space we are familiar with, into outer space, and then back to the space of an atom left nobody unaffected. The relativity of scale inspired a whole new set of research, projects and discoveries. In 1968 Constantinos Doxiadis's book *Ecumenopolis:*[13] *Tomorrow's City* was proclaimed Britannica Book of the Year. Greek architect and town planner Doxiadis collaborated closely with Buckminster Fuller and Marshall McLuhan, and they would meet regularly at the Delos Symposia,[14] organized by Doxiadis and his team. Doxiadis's vision for the future, epitomised in Ecumenopolis, alerted us to what might happen if we fail to understand the new circumstances of population growth and accelerated urbanization. He therefore proposed The City of the Future, an endless network covering the surface of the Earth, which will retain human content and anthropometry, regardless of its size. Recently Doxiadis was honoured as the first global planner.[15] In the 1970s his ideas seemed close to science fiction, but today they are surprisingly pertinent, both in content and rhetoric:

---

[12] See Buckminster Fuller, *Operating Manual for Spaceship Earth,* (1969), p. 15 Wittily, the author notices that although the manual for our spaceship is missing, we are all slowly but surely learning how to 'anticipate the consequences of an increasing number of alternative ways of extending our satisfactory survival and growth – both physical and metaphysical.'

[13] Ecumenopolis, from Greek: οἰκουμένη, meaning 'world', and πόλις meaning 'city', thus a city made of the whole world. (Source: http://en.wikipedia.org/wiki/Ecumenopolis Accessed: 02.02.2014)

[14] The Delos Symposium was an international forum discussing human settlements. The idea emerged as a reaction to CIAM meetings, but the whole thing was taken to a different level. First of all, unlike CIAM, it was not restricted to architects, but experts from all fields were invited. Second, participants did not sit in front of their finished projects but in front of an empty black board, creating rather than reflecting. Third, the symposium was not about the obvious and visible, but absolutely the opposite – the intention was to reveal invisible networks. (See: Mark Wigley, "Network Fever" in *Grey Room* 04 (2001): pp. 82–122.

[15] See: Rem Koolhaas and Hans Ulrich Obrist, *Project Japan: Metabolism Talks* (2011)

> *Today, more and more cities are not only becoming dynamic, but continuously dynamic, so that they never stop changing and growing. For the first time in history, settlements are not only three dimensional but four dimensional, since they exist continuously within the fourth dimension of time.*[16]

## The End of Distance and the Hybrid State of Space

> *The surface of the Earth has a thickness but it could not be considered a volume. The Earth's is a surface thickened by relations. [...] It always overflows any map that tries to frame it, because there are always more connections to be made.*[17]

In Google Earth, the reduced, informational appearance of a (geographical) map is replaced with a hyper-real, graphically accurate simulation of terrain. In this saturation of content, the characteristics and beauty of the landscape, its territorial borders and the socio-political character of the territory disappear, becoming relative, and set as secondary.

In his *End of Geography* thesis, Kenichi Ohmae (1990) proclaims the end of nation-states and the old cartography, announcing the death of distance and the triumph of technology over place – the annihilation of space by technology.[18] Vincent Mosco sees this as 'a story about a fundamental change in relationship to space and place', concluding that 'cyberspace is fundamentally different from geography as we know it, because this space is almost fully transparent with respect to communication'.[19] From today's perspective, one can argue that the relation between digital (cyber) and real space is symbiotic and causal, where the former does not exclude the latter. Today, the boundary between digital and real space has blurred, producing a hybrid terrain where *the digital* relies on *the real* for its means of conveyance, and *the real* relies on *the digital* for content and experience. The impact of this overlap of the real and the digital on architecture has still not been widely addressed, causing a collective uncertainty about where digital technology fits into the architect's toolkit.

---

[16] See: Constantinos Doxiadis, *Ecumenopolis: The Inevitable City of the Future*, (New York: W.W Norton & Company, 1974), p. 7.

[17] Eyal Weizman, *Forensic Architecture: Lexicon,* source: http://www.forensic-architecture.org/lexicon/

[18] See Kenichi Ohmae, *Borderless World* (New York: Harper Collins Publishing, 1990), pp. 19–20. 'The old cartography no longer works. It has become no more than illusion.'

[19] See Vincent Mosco, *The Digital Sublime: Myth, Power, and Cyberspace* (Cambridge, MA: MIT Press, 2004), p. 86.

The case of Google Earth questions the fundamental role of architecture and its aspiration to influence the perception of the world and the shaping of society. Confronted with this phenomenon, architects are facing the inevitability of a hunt for new fields for action. For architecture in this case there is again the possibility of the expansion of scope and territories.

## Conclusions

By interpreting the appearance of linear perspective and photography as crucial shifts in perception, this paper argues that the virtual globe is a symptom of a crucial current change. Architects have acquired a new tool, which with each passing day, event or technological innovation is becoming more layered and complex. We will be able to access databases of commerce dynamics, real estate fluctuations, the motion of people and vehicles, natural disasters and much more, and to trace changes over time. Can the information overload be channelled and structured into useful knowledge? How can we understand the data we collect and how can we use it to generate tools for action?

At the same time, Google Earth offers a new set of possibilities for architects and a wider scope for their primary subject of desire: the surface (as a construction ground), giving architects a sense of expansion. Expansion in this sense raises the notion of scale with regard to new possibilities of a boundless surface. New transparency of scale, distance and topography has enabled wider scope for architecture.

The contextuality of an architectural project therefore takes on a new, more complex sense in light of the accuracy of the satellite image. The new overview of territory holds the promise of a new, accurate tool that can greatly simplify and inform the process of design. Despite this, the view from beyond is not merely a utilitarian device, but possesses the poetics of discovering new landscapes and perspectives. The territory of the planet is once more a field of exploration; using holistic optical apparatus for surfing over scales and datascapes, there are no borders which act as obstacles to this exploration.

## References

Baudrillard, Jean, *Selected Writings* (ed. Mark Poster), Stanford: Stanford University Press, 2001

Colomina, Beatriz, *Privacy and Publicity: Modern Architecture as Mass Media*, Cambridge, MA: MIT Press, 1994

Doxiadis, Constantinos, *Ecumenopolis: The Inevitable City of the Future*, New York: W.W Norton & Company, 1974

Deleuze, Gilles, *Difference and Repetition* (trans. Paul Patton), London: Athlone Press, 1994

Hinchcliffe, Tanis, "'The Synoptic View': Aerial Photographs and Twentieth-Century Planning" in *Camera Constructs: Photography, Architecture and the Modern City,* Andrew Higgott and Timothy Wray (eds), Burlington: Ashgate Publishing, 2012, pp. 135–146

Mosco, Vincent, *The Digital Sublime: Myth, Power, and Cyberspace*, Cambridge, MA: MIT Press, 2004

Latour, Bruno and Albena Yaneva, "Give me a Gun and I will Make All Buildings Move: An ANT's Viefw of Architecture" in *Explorations in Architecture: Teaching, Design, Research* Reto Geiser (ed.), Basel: Birkhäuser, 2008, pp. 80–89

Leach, Neil, *The Anaesthetics of Architecture*, Cambridge, MA: MIT Press, 2000

Ohmae, Kenichi, *The Borderless World: Power and Strategy in the Interlinked Economy* New York: Harper Collins Publishing, 1990

Wigley, Mark "Network Fever" in *Grey Room* 04 (2001): pp. 82–122

## Internet Sources

Forensic Architecture, Lexicon, stable URL, accessed: January 2, 2014, http://www.forensic-architecture.org/lexicon/surface-of-the-earth-eyal-weizman/

Air & Space Magazine, The First Photo from Space, accessed: February 3, 2014, http://www.airspacemag.com/space/the-first-photo-from-space-13721411/

# CHAPTER FIVE

# RETAIL: NEW TECHNOLOGIES AND URBAN CENTRALITY

MARTIN BRABANT[1]

**Abstract** | The urban space is located in a controversial dispute with the urban creating a link - the ground floor zone. The ground floor zone offers space for social interaction and for cultural and social lifestyles in a way of architectural design. Consumer behavior in the retail trade (the "shop" in an urban environment) is in contradiction of online trading to real shopping experience. The "binary shopping experience" (virtual) is based primarily on a quantitative comparison price. This contrasts with the "physical experience" (in real terms) to be explained skilled staff needs goods and products to take on.

The use of new technologies in a hybrid form of consumption is seen as potential. Online and mobile shopping brings new requirements to the structure of the retail sector in the urban area, resulting in the Retail (generally in "centralized locations" – of real spaces) is currently in transition. Mainly caused by a modification of the consumer and user behavior because of the strong and rapidly growing market of e-commerce in addition to the retail and the mobility behavior.

*The "retail real market " has shrunk by five percent over the previous year, 2012. The Opposite is a 27.2 percent growth in online trading.*

[1] PROJEKTKRAFT – building brands GmbH, Rheinboldtstrasse 3, 2362 Biedermannsdorf
Europe-Austria, martin.brabant@projektkraft.de and Graz University of Technology Department of Architecture Institute of Urbanism, Rechbauerstrasse 12/II, 8010-Graz Europe-Austria

*Contained therein are not food, services and online sales Marketplaces (previous commissions through Amazon or eBay for example)*[2]

**Keywords** | retail, shop architecture, social spaces, technology, hybrid forms

## Introduction

The city that the "landscape" (the non-urban areas) faces is the binding and cross-linking of a physically tangible (designed) space and the (multilayered) social and communication networks working in the background. The connected questions are about the increasingly vanishing boundaries of the centre or centres – by the expansion of space itself.

Questions of research: What is the meaning of centrality in the urban context? What is the role of retail in urban spaces? Do new technologies and multi-channelling change the “typology” of retail?

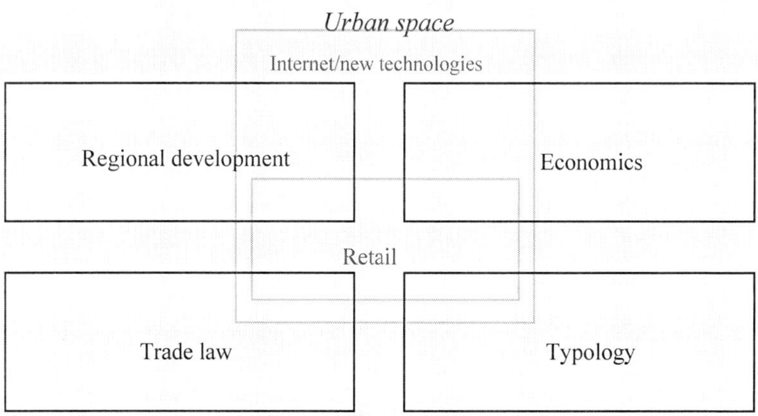


*Fig 1. Schematic operations of retail*

[2] Heinemann Gerrit: Der neue Online-Handel – Geschäftsmodelle und Kanalexzellenz im E-Commerce, Online-Handel im Wandel – Innovation im E-Commerce, Springer Fachmedien Wiesbaden (2014), p.1-2

## A The shop and its meaning

The physical urban space cannot adapt to continuous changes in society. Therefore this kind of modification is attributed to shops, since the constantly growing urban structure only allows for a limited adaption. Shops have developed from a space-building typology to a type of urban living space (anthroposphere), which has adjusted to society's needs over decades. *A shop is a product's architectonic appearance, which responds to diverse emotions (haptics, appearance, materiality, sound, aroma design) in connection with the ground-floor zone.* Nowadays, shop design means modification and identification with a special product. In selling, architecture plays the key role in which the product itself is a subordinate element. This variety of consumer behaviour, virtual product placement, social media, and technologies used in sales (rooms) need to be adapted to consumers' needs. Retail meets economically controlled mechanisms of urban structures. In an urban context a shop is a perceptible space in a cultural and social form.

## B Ground-floor zone and its meaning for central urban structures

In the Egyptian primary sector, which was the "beginning of service" (mixed farming), society and trade actually used to be based on planned economy. The explanation for food surplus is to provide food to all people. Areas of trade and barter (the modern forerunner of service and trade) gain supra-regional importance[3]. The only reason for the development of that kind of central areas is to satisfy the society's basic needs[4]. By interaction, and offering goods in baskets, those areas become spaces of primary order.

Nowadays there is a "revival" of demand for such areas in urban regions. Exchanging goods and goods trade offer various opportunities varying from consumption, recreational activities to social interaction. Usually shops face public space. *Places with a higher flow of customers are developed to form a central contact point.* Due to centrality (in the form of offer and demand), space is created.

When restructuring bartering to a monetary-based goods' relationship, there are divergent results to urban space. The ground-floor zone is formed

---

[3] cf. Reiz M.: Alltag im alten Ägypten , Augsburg (1999), p.15

[4] cf. Gutgesell M.: Wirtschaft und Handel, in Schulz R. und Seidel M. (Hg.): Ägypten – Die Welt der Pharaonen (2004), p.14-15

by urbanising through services (secondary sector) and in addition, a new kind of consumption is created. As a result of the constantly growing significance of monetary value (related to goods and services) ground-floor zones occupy positions of power in society. Spaces become an instrument of demonstrating power caused by their geographical position (central location) and resulting competition. *Centrality with supra-regional importance caused by investments has immediate effects on urban space.* Impacts are related to the ground-floor zone whose wide range of goods characterises public space. There is a relationship between a mercantilistic governmental structure and development of central spaces. Ground-floor zones (market and fair stands) are controlled and managed by authoritarian leaderships[5]. Market and fair stands (ground-floor zones) are easily affected by these and assume a separate existence (markets gain in significance). Within this process, an awareness of location (image) is included. A constant rise in commodity trading results in new typologies (places of sale, stock and storage sites), which become a characteristic for prestigious buildings, industrial zones and trade districts. With regard to ground-floor zones, this means a permanent settlement made up of craftsmen (who stopped offering goods on markets). Consequently, urban space is altered and built.

Fluctuation (similar to a "declining" cityscape nowadays, which is dominated by vacancies) is not only an expression of modernism. Additionally, it has an important impact on trading hubs' locations. The trading centres' hierarchies arise from infrastructural advantages of their location as well as from (inter)national connections and urban living that results from this (somewhat similar to Christaller's theory of central areas). Therefore centrality (referring to towns but on a smaller scale) achieves another level of social importance within urban neighbourhoods. The component of long-distance trading (caused by economisation) makes it possible to expand the trade lane from which a new net of infrastructure and (supra-)regional spaces is created. In return, the creation of such spaces leads to a new kind of typologies (due to topographic, local and structural facts of an area). *The ground-floor zone is a kind of backbone of new, growing areas and tries to meet the requirements of trading and transportation of goods. Those requirements are caused by an international net due to the location of a commodity trading hub based on a purely functional basis (for further infrastructure and supply).*

[5] cf. Foucault M.: Kritik des Regierens – Schriften zur Politik, Die „Gouvernementalität (Vortrag) Vorlesung vom 1. Februar 1978, Suhrkamp Verlag Berlin (2010), p.107-108

The functional typology (quick supply of travellers) is characterised by building a ground-floor salesroom, above which goods are stored upstairs on the top floor (new desire of goods caused by fluctuation). Market places are created by temporarily post-constructions and depend on time (highest attendance). Furthermore, these places gain supra-regional importance (at particular times). As a result of permanent settlement of traders a new wealthy social class emerges which is trying to express its wealth with magnificent houses in addition to doing justice to a stable urban location. Materiality expresses social wealth because of location (being a centre of trade)[6]. Seen from a centralistic point of view, cities play a significant role in location caused by their appearance. In addition, cities become powerful places because of the self-expression of bourgeoisie. Based on their architecture, places develop to become cities and representative areas. The following spatial developments are fundamental for categorisation. Apparently thoughtless accumulations of areas appeared which will never fulfil any terms that characterise a city[7]. These observations are important for a categorisation of distribution by size, frequency, areas and population (but nowadays in postmodern cities they are not up to date as a result of differing developments). *New space-building functions in respect of ground-floor zones result from industrialisation*. Modifications from individual crafting structures to standardised mass products lead to changes in the urban function of trade and commodity structure (products which used to be crafted). Furthermore, a legal base for free trading was created after relaxing absolutistic leadership and lordship. *Changes in society demand more efficient ground-floor zone typologies*. As the major part of population was used to self-supply, ground-floor zones changed to service zones of urban living whose society depends on external supply[8]. An extensive net of central supply institutions is used, which causes an increase in the centralisation of locations of urban structures. The city as a central place for services.

---

[6] cf. Kröger L., Peterek J., Pohl B.: Handel, Märkte und Wahrenverkehr. Ein historisch-archäologischer Ausblick , in Kenzler H., Ericsson I. (2006), p.4-10

[7] cf. Christaller W.: Die zentralen Orte in Süddeutschland: e. ökonom.- geograph. Unters. Über d. Gesetzmäßigkeit d. Verbreitung u. Entwicklung d. Siedlungen mit städt. Funktion, 3., unveränd. Aufl., Wissenschaftliche Buchgesellschaft Darmstadt (1980), p.11

[8] cf. Spiekermann U.: Basis der Konsumgesellschaft: Entstehung und Entwicklung des modernen Kleinhandels in Deutschland 1850-1914, Beck´sche Verlagsbuchhandlung, München (1999), p.84-86

The revolution of ground-floor zones can be seen as a measure for potential urban growth, which clarifies the size of a place in proportion to the people living there and trade. The bigger a place, the bigger the rise in turnover. This means, according to Christaller's development of a central structure, that the more centrally a place is located to goods, traffic and population, the more important its meaning is compared to others (regions).

On average in 1882 there were seven times more people living in cities who were employed than before. Spiekermann explains the concentration in trade (meaning of ground-floor zone) caused by industrialisation with the following criteria: *A Relative employment B The city's existence C High importance of centrality (importance of trade) D Trade-intensive settlements (relation between urbanity and labour force participation)*[9].

> *Concentration of services increases with the centrality of a place and does not depend on a city's structure.*
>
> *An average (medieval) city which is planned for approx. 2000 inhabitants cannot withstand the development and therefore has no significant importance for centrality*[10].

Those locations offer space to build the necessary infrastructure (train stations) whose centrality leads to additional changes in urban and space development, which in turn has impacts on the ground-floor zone[11]. A new type of centrality based on urban and site development is oriented toward society's needs. The character of centrality is changed by the offensive to acquire space and an industrial shaping of the city. Spatial and functional separation of the functional area is an aim.

---

[9] Schöller (1962) im allg. Vergleich von Spiekermann U.: Basis der Konsumgesellschaft: Entstehung und Entwicklung des modernen Kleinhandels in Deutschland 1850-1914, Beck´sche Verlagsbuchhandlung, München (1999), p.97-98

[10] Flachenecker H.: Die Städte im Hochstift Eichstätt während des Spätmittelalters , München, 1999; p.178f

[11] cf. Delfante Ch.: Architekturgeschichte der Stadt – Von Babylon bis Brasilia, Primus Verlag Darmstadt (1999), p.162

## C Spatial separation, which is enforced by industrial-technical instruments (production and infrastructure), leads to new urban lifestyles

In this context, it is important that the city undergoes a rationalisation process, which means using urban space in the most efficient way and orientating it to the requirements, resulting in a new spatial modulation (industry) with essential needs (main function): A Supply B Relaxation C Living D Education

This leads to a new typology of the ground-floor zone. To retain the image of the "city as a feature of modernism", commodities (processed goods and raw material) need to be organised and distributed in the best possible way. The physical distance causes a centralised site for selling goods. This is the "hour of birth" of huge stores and avant-garde shopping malls[12]. *This type describes mass consumption of goods (caused by industrial production) of modern society. Goods are a necessary resource for supplying a city's inhabitants.* Production, production plants and their goods are worthless without the establishment of a ground-floor zone, which is necessary for sale and the modern consumer society. As profit which needs to be made cannot be achieved without sales, the level of industrialisation is measured by the ground-floor zone. Shop sales should be achieved by a new kind of extensive intermediate trade. Furthermore, shops contribute value to a product and are important for a product's image (images create social bonding).

*A shop, place (social living) where a product (not work) is the focus is a place of consumption (economical space).*

Products are actually bought at shops and not in production plants or factories. However, the ground-floor zone stands for numerous possibilities of goods and purchase as well as industrialisation. As shops are kind of the "basic information" of a space-building element, development of shopping malls is based on a shop's small sections. While ground-floor zones are architectonically subordinate to production plants, (department) stores occupy a role in transporting values (an interim solution of the secondary sector [industrial sector] to the tertiary sector

---

12 cf. Fürts F., Himmelbach U., Potz P.: Leitbilder der räumlichen Stadtentwicklung im 20. Jahrhundert – Wege zur Nachhaltigkeit?, Bericht aus dem Institut für Raumplanung 41, Universität Dortmund – Fakultät Raumplanung, Dortmund (1999), p.23

[services])[13]. A change in social structures and living is caused by growing trade (more space within ground-floor zones is needed). So the shop is the focus of urban living (a few hours ago, it was producer of a "worthless" product – now it is a consumer of goods). Within urban space, trade (lifeless goods) is the focus of communication (for a few minutes at the centre of attention within a passing urban life). *The ground-floor zone is a space which is subjected to changes in society and represents the centre of urban life*. A change of a ground-floor zone to (department) stores is in conflict with centrality. The spreading of the ground-floor zone over urban space helps unstructured space to restructure. As a result, the urban body starts to consolidate (urban space)[14].

The city manifests itself as an urban space playing a major part within regional systems. Within this context, a result is the implementation of intra-urban (department) stores. Therefore there is competition between this central agglomeration of goods and typologies used in sale so far. Furthermore, this typology used within the town is more competitive than decentralised stores in the suburbs.

The urban structure's stabilisation is reached through the ground-floor zone (which represents itself as a role model for modern cities)[15]. From an architectural point of view, ground-floor zones open up, which results in a certain "sale typology" based on demand and curiosity (displaying of goods). This kind of typology faces an open space (street) and allows interaction between goods and passers-by, making the streets a place of social interaction. *Centrality as an architectonical form and expression communicates materiality and design. Shops and their consistent increase have impacts on design (appearance) of space and, of course, on social living within the city, resulting in cities' concentration and focusing*. Due to industrialisation, stores are a new kind of typology of consumer society. The massive appearance of stores (in contrast to individual shops) within the city symbolises the distance to customers. Products are now changing to "branded articles" representing and satisfying individual customer's needs instead of standardised mass products. This differentiation needs to

---

[13] cf Spiekermann U.: Basis der Konsumgesellschaft: Entstehung und Entwicklung des modernen Kleinhandels in Deutschland 1850-1914, Beck´sche Verlagsbuchhandlung

[14] cf. Allgemeine deutsche real-encyklopädie für die gebildeten stände: (Conversations-lexikon) Google eBook, Fleischhauer und Spohn (1831), p.539

[15] cf. Spiekermann U.: Basis der Konsumgesellschaft: Entstehung und Entwicklung des modernen Kleinhandels in Deutschland 1850-1914, Beck´sche Verlagsbuchhandlung, München (1999), p.141-142

be reflected in production, customer orientation as well as in sales. Department stores offer spacious areas for presenting goods and products. Consequently, product placement gains more and more in importance for customers. Products sold at shops located in ground-floor zones are offered by the sellers themselves. Standardised mass products need to be explained (that is why the same product fits everybody). By contrast, in department stores products are shown and demonstrated to clients, individually focusing on their needs. The product's individual character speaks for itself and is not subjected to an ordinary good's presentation. *Self-service (as a forerunner of trade nowadays) allows for more flexibility and possibilities for customers. They have the choice of checking more than just one product (the resulting typology of the centralised store within the city that has existed until now).* The pooled location of stores results in new micro and macro centralities. The customers' rush demand on stores for branded articles (without consulting) to a central contact point starts to soften the urban body in respect of ground-floor zones. Consequently, ground-floor zones are a declining urban typology. The widely spread centrality is occasionally concentrated in huge stores. A city's sphere which is modified by the ground-floor zone is strong enough to oppose those changes. Centrality meets social requirements and embodies urbanity! As a consequence of customers' self-determination (which product to buy), there are many resulting advantages[16]. The time which was used for a hybrid of client consulting and informal consumption is replaced and is now available. The informal communication within the ground-floor zone (classical shop) needs to be freshened up with a kind of typology which allows stays (places, spaces which invite you to linger).

Ground-floor zones and stores' dynamic development was interrupted by the Second World War and growing European trade, and is experiencing new social influences. A new type of self-service takes up space in shops situated in ground-floor zones based on face-to-face contact and stores which were partly self-serviced. An anonymous design of goods' presentation leads to efficiency (a widely-spread structure of ground-floor zones – local supplier, but slight decrease in centrality of stores).

Cities which were destroyed by the War need an extensive support of staples. Ground-floor zones offer enough space for storing as well as selling goods and are as efficient as bigger houses (in respect of stream of people). Due to quick supply, social and military influences, a new type of

---

[16] cf.. Wittl S., Häberle Ch.: Die Ursprünge des Displays von der Antike bis zur Selbstbedienung, (2010) p.19-20

trade is created (arising from basic needs). With the efficiency of self-service, which takes an important role in trade, diverse goods can be offered concentrated on one site. Finally, standardisation of products leads to the success of self-service. As packaging needs to attract customers, distribution is no longer linked to the communication with clients. This kind of "new" efficient system demands a new typology (ground-floor zone)[17]. Rather similar to (department) stores, supermarkets (as new typology) within the city offer various goods. However, in respect of space, supermarkets are managed in a more efficient way because of their flat area. The availability of a variety of goods (ground-floor zones) and offers creates new characteristics of urban living[18]. In order for urban areas not to fall apart (as this system expands faster than urban areas), there is centralisation to one economic system, whose impacts are inspired by regional development (spatial and structural specifications). Centrality is now defined by social requirements and needs (urban space falls down on this job). *Centrality loses importance because the understanding of trade cannot connect to urban space fast enough.* As time has passed, several new types of supermarkets have been developed. Based on social changes (affluent society which changed completely to external supply) even new types of trade (sale) emerged. Due to the increase in goods' specifications, supermarkets' typologies mutated to different subtypes. Owing to these specifications which result from the liberalisation of trade and sales, more and more services arise. In return, the ground-floor zone takes the role of satisfying all these kinds of needs. As a result, urban appearance is not only "controlled" by local suppliers (of existing ground-floor zones) any longer. Ground-floor zones are no longer the focus of urban living ("antiquated" social desires) and need to adapt to changes by using space for leisure, entertainment and excitement. Consequently, supermarkets are modified to specialty stores, self-service stores, hypermarkets and discounted stores and so on. Concentration on "branded articles" based on the desire to maintain a close customer relationship means that typologies "follow" their target group (settlement development in suburbs). *Products are targeting their customers.* As a result, a decentralised development of new typologies appears. As these types have economic advantages over "small" ground-floor zones in respect of area and volume, this leads to peripheral siting[19]. When such typologies shift to the tertiary sector (kind

[17] cf. Hallier B.: Die Entwicklung des POS-Marketings, (2010) p.28

[18] cf. Banken R.: Schneller Strukturwandel trotz Stabilität. Die Entwicklung des deutschen Einzelhandels 1949-2000, (2007), p.117-119

[19] cf. Wittl S., Häberle Ch.: Die Ursprünge des Displays von der Antike bis zur Selbstbedienung, (2010), p.23

of breaking urban space), they are responsible for disparity. Centrality loses significance for social needs as well as for planning theories and practices. Urban structures are morphed at arbitrary places for creating and forcing centrality. This concentrated form of typology (of consumption and sale) tries to "control" space. *A decentralised transformation of centrality to any location fails because the city's borders have no spatial reference*. Furthermore, "shopping typology" tries to force a concentrated urban structure and its objectivity was not brought into question at all. Centrality was not considered in terms of changes in society and in addition, mobility, working and leisure behaviour changed dramatically with the affluent consumer society. *Centrality in an urban context loses its quality of improving the urban living environment*. Centrality as a method of planning and development is replaced by shopping malls (which are basically central but cannot be strengthened by any space-building element). Moreover ground-floor zones are rationalised to urban supply. In the end, increasing duality (disperse spaces instead of compact city) centrality over ground-floor zones causes dissolution of urban space.

## D Modern form of retail landscape

Retail is contradictory to spatial development. Spatial development refers to virtual as well as to real goods' presence. Nowadays shops are asked to fulfil a variety of requirements (online and stationarily). In respect of ground-floor zones, shops assume a central function (control element). Shops are subject to private as well as public interest. On the one hand, in respect of earning profit as a company and on the other hand, of controlling and obtaining urban structures (decoupled function). For retail, ground-floor zones mainly depend on footfall (people entering a shop) and "urban living" is characterised by that (peoples' flow within urban space). This kind of duality is not contradictory at all. The percentage of online trade (online presence) is increasing in contrast to stationary trade. Here, stationary trade is required to adapt itself as well as possibly to distributing and operating systems and to accelerate with the help of new technologies. Additionally, new hybrid offers are needed to keep the flow of customers to "central" places for supporting and strengthening urban structures and retail.

Consumer behaviour emphasis is mainly on stationary (61%) and online stores (31%) whereby the remaining percentage is proportioned to catalogues and mobile stores (4% each)[20]. Here, the attention is directed to

[20] cf. ATKearney: Recasting the Retail Store in Today´s Omnichannel World (2013) p.3

a hybrid form – multi-channel trade is challenged with the task of finding a balanced way of combining stationary trade and online shopping. This causes a variety of requirements that retail needs to live up to. *Ground-floor zone which is changing to a multiple zone of different requirements (social consumption, freedom, adventure) hardly satisfies all these requirements.*

Usage of new technologies like smartphones and tablets is modifying retail known so far. Research is carried out by customers themselves (target-group specific offers) on the web.

> *"Customers appreciate advantages of a mobile, interconnected life and have expectations to their purchase. Site-specific services like "kaufDA" allow stationary retailers to inform (prospective) clients on the net wherever they are."*[21]

Consequently, retailers are challenged with mobility and distribution costs. Because of deliveries to decentralised places and smaller amounts, B2C causes higher costs than B2B. Equally, implementation of a central point of "will collect" is expensive as there is no advantage for the consumers. Although package stations have great potential, they are limited to size and do not offer specific attractions. One of many reasons why systems based on an individual collection of customers are successful are inflexible opening hours[22]. In a "new economy" characterised by flexible working hours and workplaces, standardised opening hours cannot satisfy customers in the long term (a customised service for users is needed).

Within a urban context, this means bunching interest and offers and increasing attractiveness. The results of the spreading of disperse places are that such individual offers (opening hours) cannot be realised as individual (private) effort in respect of travel time would be too large to use time effectively. Using new technologies in retail is groundbreaking even in other business activities. Due to individual needs, new ideas are thought up and lead to formations of new urban lifestyles.

## Conclusions

There is a variety of expectations and requirements that retail should meet. Furthermore, retail's functions are diverse, from supply to maintaining and

---

[21] Heinemann Gerrit: Studie: Mobiles Internet als Rettungsanker für den stationären Handel (2013), p.2

[22] cf. AT-Kearny: Von B2C zu B2B durch alternative Zustelloptionen – Aktuelle Herausforderung für den Paketdienst im B2S-Segemnt (2012), p.2-6

controlling urban structures to creating space and responsibility for a city's image. Shops are responsible for architectonical design and physical perception of space as well as supply in the form of a complex sectoral mix (maintaining and controlling urban structures with the help of Strassafine and target-group specific offers.) *New space is created by new requirements with the use of new technologies*. Finally, the connection between a central-space solution and new urban technologies can be controlled by ground-floor zones (shops).

## References

Allgemeine deutsche real-encyklopädie für die gebildeten Stände: (1831) (Conversations-lexikon) Google eBook, Fleischhauer und Spohn, url: http://books.google.at/books?id=qhw2AAAAMAAJ&printsec=frontcover&hl=de &source=gbs_ge_summary_r&cad=0#v=snippet&q=urbanit%C3%A4t&f=false at 23/10/2013

AT-Kearny: (2012) Von B2C zu B2B durch alternative Zustelloptionen – Aktuelle Herausforderung für den Paketdienst im B2S-Segemnt

ATKearney: (2013) Recasting the Retail Store in Today´s Omnichannel World

Banken Ralf: (2007) Schneller Strukturwandel trotz Stabilität. Die Entwicklung des deutschen Einzelhandels 1949-2000, Hg. in Jahrbuch der Wirtschaftsgeschichte von Reckendress A.: Die bundesdeutsche Massenkonsumgesellschaft 1950-2000, Akademie Verlag GmbH, Berlin

Christaller Walter: (1980) Die zentralen Orte in Süddeutschland: e. ökonom.-geograph. Unters. Über d. Gesetzmäßigkeit d. Verbreitung u. Entwicklung d. Siedlungen mit städt. Funktion, 3., unveränd. Aufl., Wissenschaftliche Buchgesellschaft Darmstadt

Delfante Charles: (1999) Architekturgeschichte der Stadt – Von Babylon bis Brasilia, Primus Verlag Darmstadt

Flachenecker Helmut: (1999) Die Städte im Hochstift Eichstätt während des Spätmittelalters; in: Flachenecker, Helmut (Hrsg.): Städtelandschaften in Altbayern, Franken und Schwaben – Studien zum Phänomen der Kleinstädte während des Spätmittelalters und der frühen Neuzeit; München ; zit. durch Rieser J.: Die Geschichte der Industriellen Revolution – Heft 1: Die Entwicklung der Stadt im Industriezeitalter

Foucault Michel.: (1978) Kritik des Regierens – Schriften zur Politik, Die „Gouvernementalität (Vortrag) Vorlesung vom 1. Februar, Suhrkamp Verlag Berlin (2010)

Fürts Franz, Himmelbach Ursus, Potz Petra: (1999) Leitbilder der räumlichen Stadtentwicklung im 20. Jahrhundert – Wege zur Nachhaltigkeit?, Bericht aus dem

Institut für Raumplanung 41, Universität Dortmund – Fakultät Raumplanung, Dortmund

Gutgesell Manfred: (2010) Wirtschaft und Handel, in Schulz Regine und Seidel Matthias (Hg.): Ägypten – Die Welt der Pharaonen (2004), Königswinter, zit. durch Wittl S., Häberle Ch.: Die Ursprünge des Displays von der Antike bis zur Selbstbedienung, in STI-Group (Hrsg.): Greif zu und Kauf mich – Displays als Erfolgsmotor für Marken und Handel, Deutscher Fachverlag GmbH, Frankfurt am Main

Hallier Bernd: (2010) Die Entwicklung des POS-Marketings, S. 28, in STI-Group (Hrsg.): Greif zu und Kauf mich – Displays als Erfolgsmotor für Marken und Handel, Deutscher Fachverlag GmbH, Frankfurt am Main

Heinemann Gerrit: (2013) Studie: Mobiles Internet als Rettungsanker für den stationären Handel

Heinemann Gerrit: (2014) Der neue Online-Handel – Geschäftsmodelle und Kanalexzellenz im E-Commerce, Online-Handel im Wandel – Innovation im E-Commerce, Springer Fachmedien Wiesbaden

Kröger Lars, Peterek Jan., Pohl Benjamin.: (2006) Handel, Märkte und Wahrenverkehr. Ein historisch-archäologischer Ausblick , in Kenzler H., Ericsson I. (Hg.): Rückspiegel – Archäologie des Alltags in Mittelalter und früher Neuzeit, Gutenbergverlag Druckerei GmbH Leipzig

Reiz Manfred: (2010) Alltag im alten Ägypten (1999), Battenberg Augsburg zit. durch Wittl S., Häberle Ch.: Die Ursprünge des Displays von der Antike bis zur Selbstbedienung, in STI-Group (Hrsg.): Greif zu und Kauf mich – Displays als Erfolgsmotor für Marken und Handel, Deutscher Fachverlag GmbH, Frankfurt am Main

Schöller Peter (1962): Der Markt als Zentralitätsphänomen. Das Grundprinzip und seine Wandlungen in Zeit und Raum, Westfälische Forschungen 15, 82-95 in Basis der Konsumgesellschaft: (1999) Entstehung und Entwicklung des modernen Kleinhandels in Deutschland 1850-1914, Beck´sche Verlagsbuchhandlung, München

Spiekermann Uwe: (1999) Basis der Konsumgesellschaft: Entstehung und Entwicklung des modernen Kleinhandels in Deutschland 1850-1914, Beck´sche Verlagsbuchhandlung, München

Wittl Sophie, Häberle Christoph: (2010) Die Ursprünge des Displays von der Antike bis zur Selbstbedienung, in STI-Group (Hrsg.): Greif zu und Kauf mich – Displays als Erfolgsmotor für Marken und Handel, Deutscher Fachverlag GmbH, Frankfurt am Main

# CHAPTER SIX

# COMFORT OF OPEN PUBLIC SPACES: CASE STUDY NEW BELGRADE

ALEKSANDRA ĐUKIĆ[1], NEVENA NOVAKOVIĆ[2] AND SRĐAN STANKOVIĆ[3]

**Abstract** | Public open spaces are very important urban elements for sustainable contemporary cities as generators of different uses and different activities. The concentration of both users and activities in outdoor spaces contributes to the liveability and vitality of neighborhoods and cities. Recent research has shown that the urban microclimate is a very important factor in the intensity of use of public open spaces, as the thermal conditions of the environment affect people's behaviour. However, the microclimate comfort of public open spaces depends on urban design and morphological characteristics of the environment. Understanding the relationship between environmental conditions, human behaviour and the patterns of usage of public open spaces can contribute to better design of outdoor environments and enhance the positive social and environmental effects of the design. The purpose of this paper is to highlight the complexity of the relationship between microclimate comfort in public open spaces and its morphological and other spatial features, and to emphasize the importance of this relationship being taken into account in the context of contemporary urban design. This paper presents a case study on a residential block in New Belgrade in which the correlation between urban form and microclimate comfort has been analyzed.

---

[1] Dr. Sci., Associate Professor, University of Belgrade – Faculty of Architecture, Belgrade, Serbia, adjukic@afrodita.rcub.bg.ac.rs

[2] Dr. Sci., Senior Teaching Assistant, University of Banja Luka, Faculty of Architecture and Civil Engineering, Banja Luka, Bosnia and Herzegovina, nnovakovic@agfbl.org

[3] PhD Student, Imperial College London, London, United Kingdom, s.stankovic12@imperial.ac.uk

**Keywords** | public open space, urban microclimate, thermal comfort, user needs, New Belgrade

## Introduction: social and ecological aspects of the use of public open spaces

Although urbanity has material properties expressed through architectural and urban forms, it is not only a spatial phenomenon. Urbanity is a kind of appropriation of spatial form by the people who use the space. It is a relation between the space and people, and of people with each other within the space. The vitality of urban space, understood as the presence of people, does not imply the presence of urbanity, but it is a necessary condition for establishing the relationship between people and space. Thus, the presence of people in public open spaces is defined as one of the basic conditions for spatial and social regeneration of cities and neighborhoods (Jacobs 1961; Hillier and Hanson 1984; Whyte 1988; Carmona et al. 2003; Gehl 2004, 2006, 2011). The success of public open spaces can be measured by the frequency of pedestrian flows and the concentration of users. The co-presence of people from different social and cultural groups is particularly desirable for the establishment of urban social heterogeneity.

Recognition of the relation between the configuration of urban space and social content suggests that sociability and social communication can be supported by the design of such spaces (Jacobs 1961; Hillier and Hanson 1984; Gehl 2006). The organization of space and its physical characteristics influence the relationships between people, their activities and ideas. The spatial configuration of cities, downtown areas, neighbourhoods and their public open spaces should be seen as a spatial framework for patterns of co-presence and activities, namely the patterns of use of public space, which affect the level and type of urbanity in a positive way.

Within the progressive schools of architecture and urban planning in Europe and the USA, analytical theories and quantitative methods for examining the social performativity of urban configurations have been developed. These methodologies involve measuring, according to various scales, the characteristics of urban structure which have a role in generating urbanity, primarily understood through encounters and interactions between people. Thus, the basic indicators are density, accessibility, connectivity, visual clarity and functional diversity, the higher values of which are considered to represent the basic conditions of urbanity. These theories consider how patterns of spatial structure and organization are directing the use of space, and thus influence the mode and quality of sociability. They also influence the sense of safety, one of the most important criteria of social behaviour and the usage of space.

However, the presence of people on the streets and plazas depends on many other factors. The characteristics of the microclimate is a highly important one, as the thermal conditions of an environment affect people's behaviour (Nikolopouloua and Steemers 2003; Nikolopouloua and Lykoudis 2006; Chen and Ng 2012). In the context of the problem of global climate change, urban planners and other urban specialists have focused on the theme of urban microclimate and the thermal performativity of urban spaces. Our energy-intesive way of life has created numerous problems which constantly threaten the endurance of ecosystems on a global and local level, while additionally climate change begins to modify our perception of the living environment in terms of accepting its limitations and perusing actions to mitigate the effects. Urban areas are the most affected by climate change, and measurements have shown an increase in the temperature of densely-built urban zones, called *heat islands* (IPCC 2007; Kuttler 2012). During the last 55 years, the air temperature in the territory of Serbia has increased at a rate of over 1.4ºC/100 years (Karadžić and Mijović 2007: 45-55). Prediction models for climate change suggest an increase of temperature of about 3-4°C by the end of the 21st century in the same territory. This phenomenon will have an impact on the infrastructure, physical structure, vegetation, agriculture, water resources and health of citizens, and these newly-created circumstances will require mitigation.

Local microclimate, together with the issue of safety, is one of the most important qualities of public open spaces and is among the main factors which drive individual perceptions and assessments of a place. The microclimate of a space influences people's decisions as to whether to use a public open space and when to use it. At the same time, spatial characteristics such as form, configuration and the combination of surface materials influence the urban microclimate. The effect of microclimate comfort – especially thermal comfort – on outdoor activities and the intensity with which public open spaces are used is a complex issue, comprising both climatic and behavioral aspects (Chen and Ng 2012). In the past twenty years, climate specialists have provided us with much data on urban climatology, and there have been broad applications of urban climatology concepts in urban studies. Chen and Ng offer an overview of the very extensive research over the past decade on behavioral aspects relating to outdoor thermal comfort (Chen and Ng 2012). But still, there is a need for deeper exploration of the correlation between the spatial characteristics of public spaces and microclimate comfort, and a need for further development and application of tools for measuring the thermal performativity of urban configurations in the process of urban design and planning.

This paper deals with the relation between the physical characteristics of urban space – such as shape, configuration and combination of surface materials – and the elements of microclimate comfort, especially thermal comfort. It offers a brief overview of the basic theoretical concepts applied in urban research on this relationship and highlights the importance of these issues in the context of urban design and planning. One set of tools for measuring the thermal performativity of public open spaces was applied to the case study of a housing block in New Belgrade (Serbia), providing an example of using evidence-based and analytical urban design methodologies in the context of thermal comfort and urban climatology.

In this paper, microclimate comfort is seen as an important factor in the social sustainability of urban space at the local level, affecting vitality, liveability and urbanity. It is also important in terms of ecological sustainability at a global level, being a part of climate conditions in general. The relationship between spatial characteristics and urban sustainability is often viewed from a particular social, ecological or economic point of view. This paper highlights the link between the ecological and social aspects of urban public space, which can be simultaneously affected by the same spatial characteristics. It also explores how a positive shift in one (thermal and microclimate comfort) can create the conditions for a positive shift in the other (vitality and liveability).

## Spatial criteria for initial assessment of microclimate comfort

Research on bioclimatic comfort in public open spaces is based predominantly on studies of thermal comfort. As defined in the ASHREA standards "thermal comfort is the condition of mind that expresses satisfaction with the thermal environment and is assessed by subjective evaluation" (ANSI/ASHRAE Standard 55). Thermal comfort depends on factors affecting the increase or decrease in body temperature and on perceptions of these changes. Environmental factors include air temperature, wind speed, relative humidity, type of clothing, etc. However, thermal comfort is influenced not only by physiological responses to highly variable microclimatic parameters, but also by psychological and cultural adaptation. Giving an overview of research on the behavioural aspects of outdoor thermal comfort in the last ten years, authors Chen and Ng summarized the conclusions of these studies, which suggest that the physiological aspect alone is not sufficient to estimate thermal comfort (Chen and Ng 2012). Thermal comfort depends on various interconnected parameters influencing the psychological adjustment of thermal conditions, such as naturalness, past experience, perceived control, time of exposure, environmental stimulation and expectations (Nikolopouloua & Steemers 2003). Tools developed for assessing thermal comfort based on aspects of human behaviour include the assessment of physical, psychological and social factors. All of these are affected by objective and subjective factors of behaviour (Chen and Ng 2012, 123).

However, there is a lack of research from the spatial perspective in response to questions about the physical characteristics of an environment that cause changes in microclimatic conditions and in the behaviour of people, including patterns of use of public open spaces. Spatial characteristics such as the shape and density of urban structures, properties of surfaces and vegetation cover affect the urban microclimate. They should be adapted to local climatic conditions and cultural patterns of space use, so they can produce a sense of comfort in public open spaces. Unfortunately, this claim was empirically demonstrated after the universal application of homogenous urban and architectural patterns in different climates and cultures all over the world during the 20th century. Although individuals' age, gender, weight, activity, clothing and exposure to the sun are very important in terms of thermal comfort and thermal tolerance, every climatic region has its own specific characteristics and meanings in terms of thermal index (Nikolopouloua & Steemers 2003). Definitions of thermal comfort have to be adjusted to the local situation and cannot be applied globally.

One of the fundamental studies on urban design morphology and thermal performance is by Gideon S. Golany, in which he explores the correlation between the spatial configuration and climatic conditions of a settlement (Golany 1996). Golany points to a number of studies and sources of information provided by climate specialists, and suggests that urban designers should translate these findings into the necessary design tools. "Although the urban design profession has been evolving around theories of other disciplines as well, it was the architectural professionals of single building design who had been considering the climatic environmental essential to their design. The outdoor urban space is left without any design attention" (Golany 1996, 455). According to Golany, when he wrote the text almost two decades ago there was almost no literature on the spatial criteria of microclimate comfort at the scale of a neighborhood or city. Contemporary studies are still rare and they often depart from Golany's basic thesis, as with recent research on the relation of geometry and street orientation with urban microclimate (Shishegar 2013). In the rest of this section we summarize some of the basic spatial principles in the context of urban microclimate, as set out by Golany, which we tested using an environmental modeling tool called ENVI-met (Bruse 2010, Bruse and Fleer 1998).

Golany explores the correlation between the main climatic features of a region and its major problems and issues, describing its overall profile, the basic urban design response and the preferred urban forms (457-458). Urban design on the level of neighbourhood or city should respond to microclimatic requirements on two levels: in the selection of urban sites and at the level of urban configurations. The basic geographic features that should be considered when selecting a location in terms of thermal performativity are: altitude, wind temperature, adjacency to a body of water, prevailing wind direction, slope, potential for flooding and orientation. These geographic features greatly affect the basic microclimate profile of locations. We can say that they are not only important during the process of selecting sites for new neighbourhoods, but also in reconstructing the spatial form of pre-existing neighbourhoods.

Looking at spatial configuration from a thermal performance point of view, every section of the city must be seen not only as a specific form, but as an integral part of the city as a whole. Golany in this regard recognizes three types of historic urban form that produce a comfortable microclimate in completely different climatic regions: compact, dispersed and clustered (459). However, they need not be applied exclusively across an entire city. It is possible to combine urban forms, depending on the location of individual urban units within the urban ensemble, in such a way that their mutual relations will produce a more comfortable sensation in certain climates. For example, certain morphological forms in peripheral parts of a city can produce better ventilation in its central areas in humid and warm climates where ventilation is important for thermal comfort. Therefore, it is important to consider the microclimate characteristic that Golany called *urban thermal differentiation*, which is particularly expressed in the relation of central and peripheral parts of a city.

Finally, some of the characteristics of spatial configuration that affect thermal performativity the most, according to Golany, are the orientation and configuration of the street system, the width of the streets, the pattern of different land-use intensity, the distribution and quantity of the vegetation, and the size of public open spaces. The author emphasizes that there are no universal spatial models of the abovementioned patterns and that they need to be adapted to each particular situation. Very often, spatial models that have certain effects on thermal comfort in one climatic region can cause completely the opposite effect in another. However, what is universal is the impact of these spatial patterns on the characteristics of a microclimate and the human perception of this. Golany explained how each of these elements of spatial configuration affects the microclimate conditions. Their different forms and designs in relation to thermal comfort remains to be explored, as well as, more interestingly, their combination and different spatial arrangement in the matrix of the city. This task is particularly challenging considering the existence of many different software tools and technical tools to assess microclimate comfort.

## Case study

The research was undertaken in a famous residential block in New Belgrade called Block 21, built during the 1960s. The residential block is one of the iconic modernist structures of New Belgrade, which have always been a symbol of and testimony to the many ideological shifts in Yugoslav and Serbian society. More than 50 years of urban existence has brought numerous adjustments and changes in its economic, social and functional structure, but its modernist urban matrix composed of slabs and towers located in a large, green open space has mostly remained untouched until recently. The major morphological and functional transformations of Block 21 happened during the last two decades.

Originally, Block 21 was built as a mono-functional residential block. Urban reconstruction went through two phases during the 1990s and 2000s. New urban functions were added, such as commercial and office venues, as well as new residential structures which were quite different from existing ones in terms of urban morphology. These transformations of residential blocks in New Belgrade had a major impact on the character and intensity of use of communal spaces. The figure-ground ratio changed in favour of ground occupancy. Naturally, the one of the consequences was a change in the surface materials used, usually towards less greenery and more covered pavement. Open spaces for collective use were transposed to another type, different in size, function and patterns of accessibility.

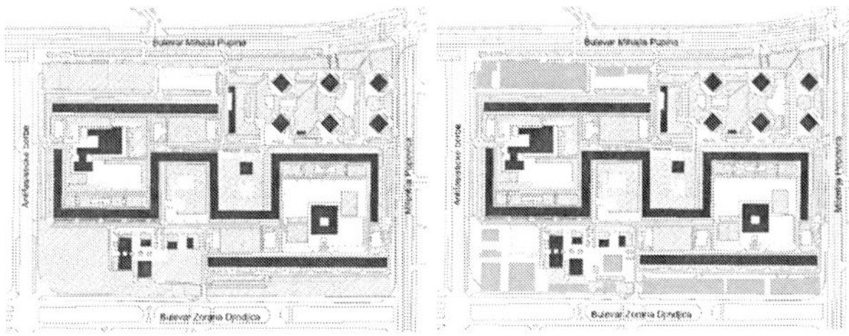

*Fig. 1. Morphological transformation of Block 21. Residential block before recent transformations (left) and after (right). New buildings marked in grey.*

## Methods

This paper is developed around the hypothesis that transformations of the spatial structures of inherited modernist residential blocks are influencing the thermal comfort of public open spaces. The case study of Block 21 aims to provide results showing the extent to which changes in thermal conditions can occur when urban structures are changed. Two research methods were applied. The first was a population survey involving personally administered questionnaires; the other was a virtual ENVI-met simulation of the thermal performativity of urban structures. At the end, the correlations between results were analysed.

The first part of the case study used the technique of personally administered questionnaires. 200 citizens, representing 2% of the population of Block 21, were questioned during two weeks in September 2011. The socio-demographic characteristics of the respondents were fairly representative of the socio-demographic structure of the inhabitants of New Belgrade and Block 21. In other words, the average age was 41.3, 48% of citizens were older than 45, and 60% had received a high school education. The aim of the population survey was to understand the users' needs regarding communal spaces inside the residential blocks by applying a system of five criteria that are related to thermal comfort: safety, accessibility, urban structure legibility, place attractiveness and place liveability.

In the second part of the case study, two different urban configurations of the same residential block were tested using ENVI-met environment modelling software. The first configuration was based on the original plan of Block 21 before the recent transformations. The second one was the spatial configuration after the transformations, which represents the present state of the block's morphology. The ENVI-met 3.1 Beta 5 model is a tool specifically developed to study and predict urban microclimates. It is a grid-based, three-dimensional model for the simulation of airflow, heat and vapour exchange, exchange at vegetation and vegetation parameters, turbulence and particle dispersion in urban areas. The software preforms micro-scale simulations with a typical resolution of 0.5 to 10 m in space and 10s in time. It takes into account key modelling inputs, including site location, initial climatic parameters, soil and plant type, building structure and thermal properties. It can generate three-dimensional distributions of radiation, temperature, heat flux, humidity, wind flow and other relevant data such as PMV (predicted mean vote). Recently, ENVI-met has been adapted mainly to simulate surface-plant-air interactions in urban canyons, and to predict the microclimatic consequences of different urban design options (Bruce and Fleer 1998).

Two files are required by ENVI-met software to run a complete simulation: the configuration file and an area input file. The configuration file includes the major settings for the simulation, such as its date and duration, as well as basic meteorological information. The area input files include information about the morphology of the environment, such as the position and height of the buildings, position and types of plants, distribution of surface materials and soil types, position of sources and receptors, the location's geographical position on earth, and so on.

Two models were thus created representing Block 21 before and after its spatial transformation, investigating the impact of new urban structures and materials and the new figure-ground ratio on the urban heat island effect. In both models the simulations were run under the same conditions. The simulation date was 24 July 2013, a day with typical summer temperature conditions for Belgrade. The initial air temperature was 19.8°C; the wind speed at 10m above ground level was 3m/s from a south-easterly direction; relative humidity up to 2m height above the ground was 50%; specific humidity at 2,500m was 7g water/kg air±1 and the adjustment factor for solar input was 1. The area input data for the ENVI-met simulation was a 100x90x30 grid; grid cell size in meters was dx=5, dy=5 and dz=3; the geographic latitude and longitude was 44.8167° N, 20.4667° E. The simulation was run for 10.00 am, 12.00 noon, 02.00 pm, 04.00 pm and 06.00 pm. The duration of the simulation was 24 hours, from 6.00 am until 06.00 am next day. Air temperature and humidity at 0.8m above the ground were extracted from the results.

## Results of the survey about users' needs

Users were asked to rank priorities concerning the everyday use of residential communal spaces. The results of the questionnaire show that residents were the most concerned about safety issues in public open spaces, especially problems arising from poor lighting. Needs concerning bioclimatic comfort were in fifth place, but 67% of residents questioned thought that their communal open spaces were not pleasant to use during the summer days. Residents were also very concerned about the level of urban equipment in public open spaces: 92% felt that there were too few parking lots and garages, while 98% felt there was insufficient green space. While poor design and low levels of equipment in public open spaces are a major problem for residents and can have a number of negative functional and social effects, they are also closely connected to the problem of thermal performativity. There is not enough greenery, according to users, and there are many more automobiles than there are parking spaces. One of the great problems of urban design and thermal comfort is how to achieve more parking lots (covered ground) and more greenery at the same time.

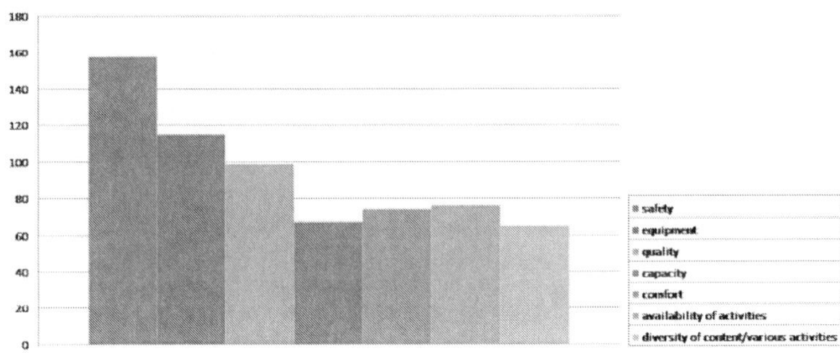


*Fig. 2. Priority ranking of users' needs in Block 21 according to questionnaire results*

Results show that communal spaces, in the opinions of the participating inhabitants of Block 21, do not fulfil certain criteria of comfort, including thermal comfort. However, the questionnaire was used to make problems and needs more visible and to direct urban designers towards adequate interventions in public spaces and the surrounding area. 80% of those questioned put the safety criteria on the first place. 55% think that general quality of pavement, urban furniture and facilities are the most important. 37% put the accessibility of open space and thermal comfort as their priority.

Although the crime rate in New Belgrade is more-or-less the same as in other parts of the city, it is interesting to observe that residents were the most concerned about safety problems. The main goal of any potential transformation of urban structures and the architecture of residential blocks should be to upgrade of the quality of the communal spaces that people use on an everyday basis, in a way that will fulfil their needs.

## Results of the ENVI-met simulation

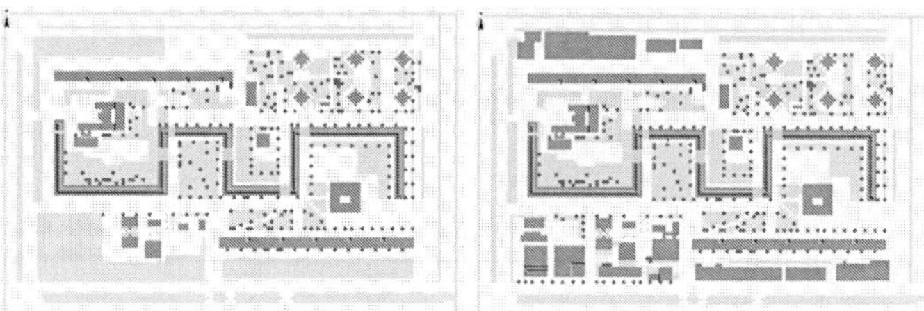

*Fig. 3. ENVI-met area model for two spatial layouts.*

Before the transformation, 54.68% of the Block 21 area was under building structures; now that percentage is 70.24. Green spaces occupied 103,172.54m$^2$ of the inner block space before the 1990s and now they cover 67,757.15m$^2$. On the other hand, the trees which were planted 40 years ago now have treetops four times bigger and create shaded areas within the inner side of the block.

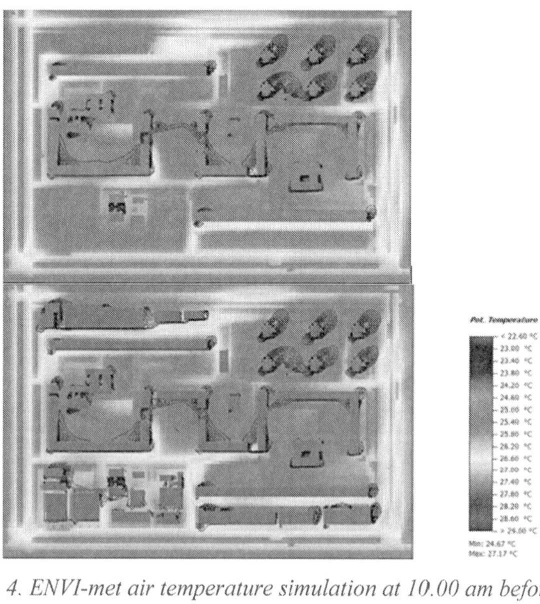


*Fig. 4. ENVI-met air temperature simulation at 10.00 am before the transformations (top) and after the transformations (bottom).*

Comparison of the outputs from the ENVI-met simulation for air temperature for the two block morphologies shows a decrease of 1 to 2°C due to shading and evapotranspiration in the central parts of the communal open space and an increase of temperature of 1°C along the edges of the block where new buildings were built. As can been seen from Figure 4, air temperatures over the parts with vegetation are much lower than those with concrete or asphalt, although some areas without vegetation still remain cool due to the shading effect. The biggest difference in air temperature reached nearly 2.5°C, comparing the two urban layouts. Also, the air temperature in the open communal spaces covered by trees was about 0.5 - 0.6 °C lower than in the areas covered by grass.

With the construction of additional buildings next to Boulevard Z, Đinđić and Antifašističke borbe Street, the temperature rose in their immediate surroundings. Comparing the air temperature simulation in the two spatial layouts, a difference can be seen at the corner of the Boulevard and Milentija Popovića Street where the temperature is 0.5 °C higher, as well as in the north part of the block in the new urban layout.

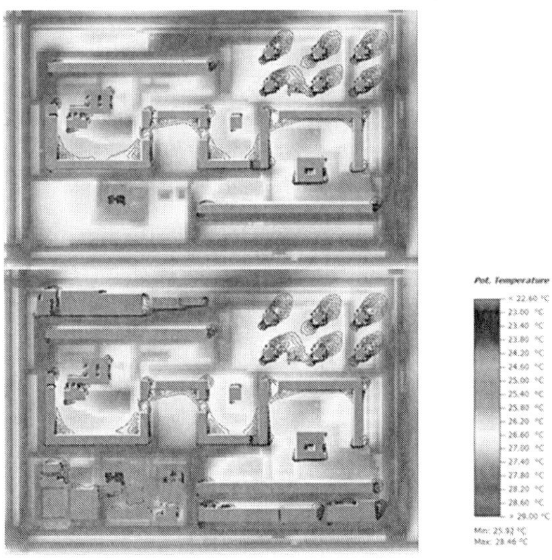


*Fig. 5. ENVI-met air temperature simulation at 14.00 pm before the transformations (top) and after the transformations (bottom).*

From Figure 5 we can get a general picture of the influence the new buildings have had on the air temperature in Block 21. As the prevailing wind is from the south-west in Belgrade, it is understandable that buildings constructed on the south side significantly changed the air temperature inside the block. The greatest increase in temperature (over 2°C) occurs next to the Boulevard where wind is blocked and hot air accumulates.

## Discussion

The investigation of residents' evaluation of Block 21's communal open spaces according the six criteria shows what is missing and what parameters an urban designer should consider during the design process. Also, this kind of survey' could be used for determining the hierarchy between parameters during the design process to best meet users' needs. According users' opinions, the main criteria in the regeneration of communal open spaces should be safety and the functional and thermal comfort of the area (more street lighting and better visual connection between places, more shade, more vegetation and more street furniture).

The percentage of green surfaces within Block 21 decreased following the transformation in the 1990s. Formerly green spaces were covered by new buildings. On the other hand, the existing trees are well dispersed across the open space of the block, and in the 40 year period since they were planted, the treetops have provided shade for most of the space next to the buildings and for flats up to the second floor. The buildings which were added during the 1990s have created a few smaller inner yards within the block which also provide more shade in the open spaces. This morphological transformation has produced a slight decrease in temperature in these newly-formed open spaces, but new covered ground to the south and north of the block has resulted in an increase in temperature along its edges. Lack of pavement maintenance and inadequate choices of materials for the pavement have also led to a slight increase in the air temperature.

Besides calculating pure meteorological data, ENVI-met also includes a simple bio-meteorological (Biomet) model to predict the thermal comfort inside the model area. The *predicted mean vote* (PMV) model is probably the best known Biomet model. It relates the energy balance of the human body to the feeling of personal comfort for a user exposed to the corresponding climates. The PMV index predicts the mean response of a large group of people according to the ASHRAE thermal sensation scale where +3 represents hot, +2 warm, +1 slightly warm, 0 neutral, -1 slightly cool, -2 cool and -3 cold. The PMV index is expressed as $PMV = (0.303e\text{-}0.036M^{+0.028})\ L$, where M stands for metabolic rate and L for thermal load, defined as the difference between internal heat production and heat loss to the actual environment (Ostergaard et al. 1974).

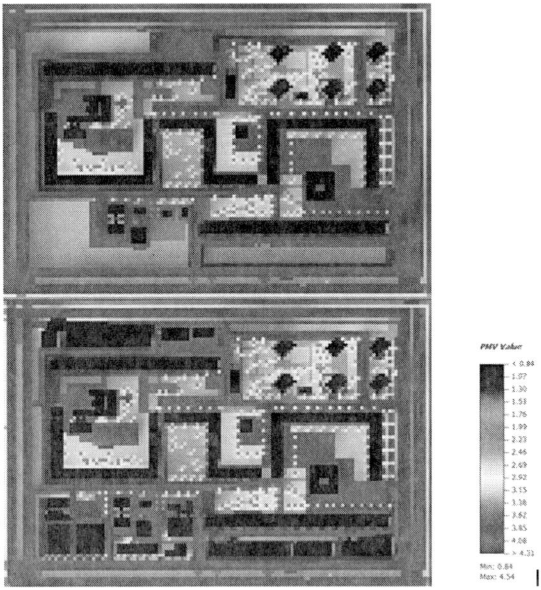


*Fig. 6. ENVI-met PMV simulation at 14.00 pm before the transformations (top) and after the transformations (bottom) .*

Analysing thermal comfort in terms of the PMV index in Block 21 produced the results shown in Figure 6, indicating that the level of comfort goes beyond +3 (hot) and the average is between +2.9 and +4. The construction of the new buildings resulted in the PMV value increasing to more than +3.5.

This study has shown that the urban planners' choice of new neighbourhood morphology had a negative impact on the urban environment and thermal comfort. As wind is one of the most important factors in temperature flux, changes to the block's configuration and form which blocked the prevailing wind have resulted in an increase in the air temperature inside the block. The maximum air temperature increase in Block 21, which can be attributed to the existence of these new urban structures, was 2.4°C at around 2:00 pm. As air temperature is reciprocal to air quality, it is reasonable to assume that with the rise in air temperature, air quality has dropped. The results of the ENVI-met simulations confirmed Golany's theory regarding the correlation between urban morphology and the thermal performativity of urban structures. They demonstrate how the density of urban structures has a relative value and that it is necessary to choose an appropriate configuration of urban density if social, functional and microclimate requirements are to be fulfilled at the same time.

## Conclusion

The significance of communal open spaces in residential blocks should be reinstated: they represent important gathering points and places for recreation and leisure, while at the same time representing various patterns of urban culture. The choice of spatial configuration, construction and surface materials, and the layout of green spaces should lead to improvement in the urban microclimate and the thermal comfort of public spaces. It should be considered essential for the regeneration of existing neighbourhoods and their public open spaces to be informed by bioclimatic parameters, if such spaces are to fulfil their role as the main place for social interactions, especially during the hot summer period.

The research undertaken for this study has confirmed that the microclimate of outdoor public spaces can be greatly influenced by the size and shape of the space, the amount of vegetation and the paving materials used, as well as its age and maintenance. It has been shown that smaller open spaces with a significant amount of vegetation can lead to a better-balanced microclimate during the summer days. Even the slightest changes in the figure-ground ratio, as well as in the size of green spaces and type of vegetation, can influence microclimatic conditions. The purpose of this paper was to highlight the complexity of the relationship between microclimate comfort in public open spaces and its morphological and other spatial features and to emphasize the importance of taking this relationship into account in the context of contemporary urban design.

## Acknowledgment

The study was done within research projects TR36035: Spatial, Ecological, Energy and Social Aspects of Urban Development and Climate Change – Interactions, which is funded by Ministry of Science and Technology of the Republic of Serbia. Special thanks to PhD student Zorana Matić and BA students of the elective course “Redesigning pedestrian networks” at the University of Belgrade, Faculty of Architecture, 2011-12 and 2012-13 cohorts, who personally administered questionnaires in Blocks 21 and 22.

## References

ANSI/ASHRAE Standard 55-2013. 2013. *Thermal Environmental Conditions for Human Occupancy.* https://www.ashrae.org/resources--publications/bookstore/standard-55

Bruce, Michael and H. Fleer. 1998. "Simulating surface-plant-air interactions inside urban environments with a three-dimensional numerical model". *Environmental Modelling Software* 13: 373-384.

Bruse, Michael. 2010. *ENVI-met Homepage.* http://www.envi-met.com.

Carmona, Matthew, Tim Heath, Taner Oc and Steve Tiesdell. 2003. *Public places - urban spaces: the dimensions of urban design.* Burlington: Architectural Press.

Chen, Liang and Edward Ng. 2012. "Outdoor thermal comfort and outdoor activities: A review of research in the past decade". *Cities* 29: 118–125. doi:10.1016/j.cities.2011.08.006

Gehl, Ian, Lotte Johansen and Sojvejg Reigstad. 2006. "Close encounters between buildings". *Urban Design International* 11: 29-47.

Gehl, Ian. 2004. Towards a fine city for people: Public Spaces for Public Life. London, Copenhagen: Gehl Architects.

Gehl, Ian. 2011. *Life Between Buildings: Using Public Space.* Washington: Island Press.

Golany, Gideon S. 1996. "Urban design morphology and thermal performance". *Atmospheric Environment* 3: 455-465.

Hillier, Bill and Julienne Hanson. 1984. *The social logic of space*, $5^{th}$ edition. Cambridge University Press.

Intergovernmental Panel on Climate Change (IPCC). 2007. *Synthesis Report.* Geneva: IPCC.

Jacobs, Jane. 1992. *The Death and Life of Great American Cities.* [1961] New York: Vintage Books.

Karadžić, Branko and Mijović Ana, Eds. 2007. *Environment in Serbia: an indicator–based review.* Belgrade: Serbian Environmental Protection Agency. http://www.sepa.gov.rs/download/Environment_in_Serbia_Full.pdf

Kuttler, Wilhelm. 2012. "Climate Change on the Urban Scale – Effects and Counter-Measures in Central Europe". In *Human and Social Dimensions of Climate Change*, edited by Netra Chhetri. doi: 10.5772/50867. http://www.intechopen.com/books/human-and-social-dimensions-of-climate-change/climate-change-on-the-urban-scale-effects-and-counter-measures-in-central-europe

Nikolopouloua, Marialena and Steemers Koen. 2003. "Thermal comfort and psychological adaptation as a guide for designing urban spaces". *Energy and Building* 35: 95-101.

Nikolopouloua, Marialena and Spyros Lykoudis. 2006. "Thermal comfort in outdoor urban spaces: Analysis across different European countries". *Building and Environment* 41: 1455–1470.

Ostergaard, J., P. O. Fanger, S. Olesen and T. L. Madsen. 1974. "The effect on man's comfort of a uniform air flow from different directions". *ASHRAE Transactions* 80(2): 142-157.

Shishegar, Nastaran. 2013. "Street Design and Urban Microclimate: Analyzing the Effects of Street Geometry and Orientation on Airflow and Solar Access in Urban Canyons". *Journal of Clean Energy Technologies* 1: 52-56.

Whyte, William H. 1988. *City: Rediscovering the center*. New York: Doubleday.

# CHAPTER SEVEN

# VERTICAL PUBLIC SPACES

SORANA RADULESCU[1]

**Abstract** | The landscape of contemporary city life is changing. Since public space is traditionally understood as a horizontal surface placed between and limited by building facades, the question of introducing new dimensions for public realm gains importance. The generation of vertical public spaces as an infrastructural interface between buildings and their urban surroundings can address frequent issues of the contemporary townscape, such as creation and activation of urban life or re/appropriation of place. Is a vertical public space a valid urban option for alienating and disorienting city areas, where the users lack in engagement with their built environment? What could this new type of interface between building and open public area be in order to increase the bond between the protagonists of the urban spectacle and the canvas? Can it become an asset that helps architects and urban designers blur the boundaries of what we codify as public and private areas, intensify and diversify the urban experience, invest the public space with new meanings and lead to rewriting the city space?

The dissolution of the CIAM established new directions in understanding the city, its dynamics of public life, behavior of individuals, their modes of encounter, gathering and bonding, of recognition and distinction, all these leaving previous paths behind. The paper will start focusing on the utopic approaches of the 1960's, their evolution up to their validity nowadays, analyzing the evolution of these architectural phenomena under the aspect of the generated vertical public space. Furthermore, new contemporary approaches unfold by understanding theoretical influences of organizations such as the International Situationists a.o. While such utopic proposals offer a ground for understanding new meanings of public space, this

---

[1] Institute for Architecture Technology, Graz University of Technology. Rechbauerstraße 12/I, Graz, Austria. sradulescu@tugraz.at.

research will focus on vertical public spaces in existing situations encountered in problematic city areas, such as new interpretations of unused sites in the European cityscape.

**Keywords** | vertical public space, urban third dimension, building interface, strategy, urban utopias of the 20th century.

## Introduction

By the year 2050, two-thirds of the world's population is likely to inhabit urban settlements[2]. This accelerated concentration of humanity in cities represents not only new challenges but also numerous opportunities to rethink and improve the way in which urban life is shaped. The obsession with gaining height is not new in architecture. From an urban point of view height accents define the skyline and represent landmarks important for orientation. Contemporary planners' ambition to grow higher comes either as a response to the constraints imposed by already dense city structures or as a tool for the generation of higher urban densities. But what about the repercussions on public spaces?

The model of the compact city is regarded as the most sustainable structure[3], aimed at establishing a human scaled street network with intensive, vibrant nodes. These circumstances underlie the research for means of introducing new vectors in the compressed matrix of public realm and releasing it into the third dimension. The first part of the investigation analyses the contemporary feasibility of urban and architectural utopias emerged in the 1960s and reinterprets some relevant features for the implementation of vertical public spaces. The chosen examples are still valid nowadays and vindicate the need to "go up". The second part of the paper focuses on the strategy on how to use and transform existing urban structures for vertical public spaces.

## Why Go Up?

### The Grey Tones

A figure-ground diagram of any urban structure allows distinguishing the built mass – the positive – from the unbuilt free space – the negative. This

---

[2] "Urban Patterns for a Green Economy, Leveraging Density". UN HABITAT for a better urban future.

[3] "Urban Patterns for a Green Economy, Leveraging Density". UN HABITAT for a better urban future.

white flat background reveals the network of the city. Common sense translates it into a horizontal surface limited and shaped by the constructed contingent. Unless determined by topographic constraints, the network of streets, squares and walkways that build up the public space of cities is generally stretching out in just two directions. I argue that public space is not obligatory a horizontal one, so introducing new dimensions to its network, letting it evolve three-dimensionally seems feasible. Under the premise of condensing the urban fabric, the denser the black areas, the thinner and more fragile the white background becomes. But can there be introduced grey tones or does the figure-ground scheme have to remain radical in its representation? There are indeed urban situations that go beyond the black or white situation: uncertain terrains, urban gaps and voids, constructions whose purpose is either lost or weakened due to a variety of historical, political, social or cultural actions that have scared the image of the city. Thus it becomes important to detect not only the nodes of already high urban intensity but even more the zones that can possibly get activated. By regarding such setups as three-dimensional plots, suddenly "building", "city" and "landscape" can switch roles. These situations need to be encountered and exploited within the spectrum of grey tones. There has to be found an interface for these "deficient" components of the cityscape in order to resuscitate them into the network of public life. In the wrinkles of the urban tissue is where real potential of activation resides.

## The Nomadic Component, The Need For Situation and The Utopias

Such undefined interstitials are the interface between building and public space. In these areas of intersection public life can be impulsed. It is also the place where the nomadic (void) meets the sedentary (solid)[4]. The nomad component in people's everydayness can strongly influence their perception of the surrounding built environment. The clearest experience comes by moving through it. As argued by several artistic movements of the 20th century, walking is the practice that allows spectators of the cityscape to switch roles and become protagonists, actively involved in the great urban performance. One of the three main components that define public spaces, besides commercial activities and social interaction, is the need to move[5]. The Situationists' "dérive" is probably the most eloquent

[4] Careri, Francesco. 2002. "Walkscapes–Walking as an aesthetic practice".
[5] Gehl, Jan, and Lars Gemzøe. 2000. "New city spaces".

initiative of experiencing and (re)discovering the city in all its "grey tones" by the act of walking. The dynamic way of perceiving the built environment by moving through it is a first vital step to the implementation of urban situations with a vertical component.

The creation of situations that continually reshape people's lives was another of the basic principles evoked by the International Situationist movement. They are meant to be performances of the everydayness, held together by a playful succession of events that also imply an artistic manifestation, "concerned with either evading or confronting a commercialized art establishment coupled with an administered culture"[6]. As culture and education serve as strong engines of urban dynamics, artistic acts increasingly catalyse public life and simultaneously become the impulse of urban development. Being so tightly related to the energy of public life, cultural manifestations in public space have been the most employed tactic of rising awareness and engaging with the city. The artistic performance offers citizens the necessary freedom to evade from the patterns of imposed behaviour and discover new paradigms of urbanity. But there are several more forgotten or ignored uses in the cityscape capable of becoming real activators of urban life. New meanings of public realm can be generated by exploiting the potential of these uses through a process of creating "situations".

The urban and architectural utopias of the 1960s still offer valid ideas for our present challenges regarding the formalization of public realm. These bold propositions rethink the city and change the paradigm of public space use. The field of study is vast but I will concentrate on famous proposals that condense several features relevant for this work: Yona Friedman's "Ville Spatiale", Constant Nieuwenhuys' "New Babylon" and Cedric Price's "Fun Palace".

Ville Spatiale remains in its representation highly utopic, but the fundamental concepts behind it can be regarded as feasible nowadays. Overlapping an existent structure with a new urban pattern condenses the city and frees the street level for public use, diminishing the impact of the built mass on ground floor. High flexibility in designing individual habitats and in deciding the grid of uses converts "Ville Spatiale" into a model of an adaptive urban system, where individual choices and free will prevail. "It can be anything (...) the shape is unimportant"[7]. One of the

---

[6] Erickson, Jon. 1992."The Spectacle of the Anti-Spectacle: Happenings and the Situationist International".

[7] Iaac Closing Lecture: Yona Friedman Interview. 2011. Accessed February 10, 2014.

topics he brought to discussion in the 1960s criticizing the endeavour to build more and more, ignoring the potential of the existing built contingent is still a valid argument. Under the premise of condensing and concentrating the city fabric, the strategy can no longer be one of constructing new spaces for new uses, but turning to the existent ones and reinterpreting them instead. Comparing the built city to hardware and its inhabitants to software, Friedman focuses on the software. Any public space can be transformed only by changing its use patterns. Such as Careri built his discourse on the relevance of the nomadic component of urban behavior, Friedman's software flows mark the pulse of public space. Still, Ville Spatiale remains utopical. Its radical output and unresolved connection between the lifted structure and the existing grounded city proves the impossibility of its implementation. Still, Ville Spatiale is an important "seed" of the third dimension of urbanity that can be developed further. Instead of lifting the spatial structure another strategy supporting the idea of a vertical public space arises: prolonging the public realm from the ground floor, where the interconnected field of public life is naturally active, in the vertical direction. Under the same logic of the adaptive system, instead of a radical superimposed structure detached from the actual edifice, a more realistic intervention would consist in merging the existing built structure with a new system that fills its gaps and develops, such as buildings, vertically.

*Fig. 1 Drawing of an example of Ville Spatiale, Yona Friedman.*

Yona Friedman sees the city as a responsive environment full of modifiable elements that bend to the will and unpredictability of its inhabitants. Constant Nieuwenhuys renders his "New Babylon" through a similar approach, basing his new city concept on the nomadic character of human behavior. There are several common features between "Ville Spatiale" and "New Babylon", such as the proposal of the lifted structure, the idea of detachment from ground level and the development of an independent structure over an existing one, the freedom of configuration of the new environment and its adaptation to the needs of the inhabitants. The major difference consists in the way of experiencing these new urban systems. Constructing on the Situationists' "derive", the act of walking is rendered again as the intrinsic nomad character of humans, often repressed in contemporary urban configurations. This study on vertical public spaces relies on the premise that they need to be encountered, generated and experienced through a dynamic act of walking. It is how the inhabitants of an urban settlement can become more aware of the immediate surroundings and have the revelation of new types of public space. It is how pedestrians take possession of the built landscape and appropriate it in personalized way.

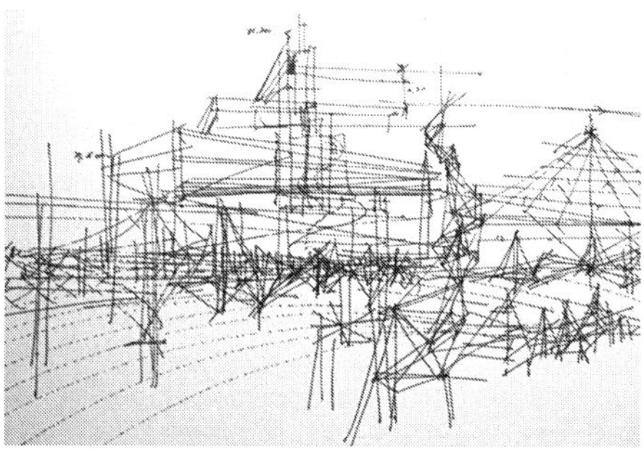

*Fig. 2 Sketch of a Sector of New Babylon, Constant Nieuwenhuys*

Ceric Price's "Fun Palace" offers a significant response to the concept of adaptive living environment. It addresses similar topics as Friedman's and Nieuwenhuys' urban utopias at an architectural scale. Fun Palace is a feasible project. Price reduces the architectural intervention to a symbolic scaffold for the movable programmatic clusters, a "short-life conglomerate of disparate, free-choice, free-time, voluntary activities"[8]. This flexible structure enables users to mould their environment in response to their personal needs, "as a site of change and impermanence"[9]. Such an intervention, implemented in the urban fabric, has the ability to catalyse the flows and variables of the city regarding the ever-changeable urban dynamics. Price introduces the notion of temporary programs that can

[8] Mathews, Stanley, Cedric Price - From "Brain Drain" to the "Knowledge Economy", Accessed February 03, 2014. http://www.audacity.org/SM-26-11-07-01.htm.

[9] Mathews, Stanley, Cedric Price - From "Brain Drain" to the "Knowledge Economy", Accessed February 03, 2014. http://www.audacity.org/SM-26-11-07-01.htm.

emerge and vanish along with the needs of the users. The filiform framework supports the dynamic unfolding of public activities both horizontally and vertically. It is a three dimensional scaffold for public space. The built mass and its impact are reduced to a minimum and the effect on public realm can be remarkable. The Fun Palace represents a tribute to the culture of the ephemeral, a huge machine stirred entirely by creative forces and also a great inspiration for contemporary architectural thinking.

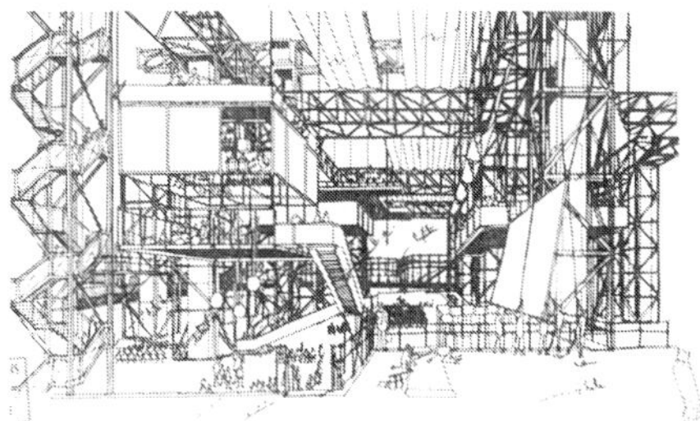

*Fig. 3 Section Drawing of Fun Palace, Cedric Price.*

## How To Go Up?

### The Strategy

"(Re)designing the city is less composition and much more improvisation. And in improvisation it is crucial to leave space for the other players."[10]

The mentioned influences of our architectural and urban recent past highlight the important aspects that reflect on the strategy of implementation of vertical public spaces: the need to experience public spaces by walking, the importance of a framework that allows situations to happen, the need to fill in urban gaps with overlapped, adaptive systems before building anything new.

There are certain steps that frame the urban improvisation and lead to the enhancement of vertical public spaces.

### 1. Discover!

The first step constitutes the identification of appropriate "sites", places within the urban tissue that are capable of opening up to the matrix of the

[10] http://www.smarturb.org/index.php?#article:110, Time Based City.

city and become an active part of public life. There are two different types of “sites”: existing urban (infra)structures that allow the addition of uses and forgotten gaps of the city that can be reintroduced in the active network of public realm. The series of detailed drawings entitled “The Happiness Machine”, by Mark Lascelles Thornton[11], provides an interesting insight on how to encounter the needed sites for creating urban situations. Picturing commonly known urban images, he highlights the “neutral” elements of the cityscape. By looking at those parts only, the viewer would not recognize the image of city. But it is exactly there, in the wrinkles of the urban tissue, in its undefined character and vague specificity that real potential of generating public identity in the vertical plane lays.

*Fig. 4 The Happiness Machine, Mark Lascelles Thornton.*

## 2. Impulse!

After having found these potential sites, public awareness has to be generated. Opening up to the common eye, the sites must become visible and attractive. The element of surprise delivers adequate tools to reveal the potentially new integrant of public space. Unexpected interventions surprise the passengers and establish the site on the city’s mental map. For example, Spanish designers of “Espai MGR” use this tool in their unbuilt proposals to reveal forgotten corners and voids of the Valencia built structure, returning these forgotten places to public awareness[12].

[11] http://marklascellesthornton.blogspot.co.at. Accessed 01.02.2014

[12] http://www.espaimgr.com/habits-make-us-blind/. Accessed 01.02.2014

*Fig. 5 Visualisation, Espai MGR.*

## 3. Metamorphose!

The element of surprise vanishes over time and the intervention can be reduced to a mere and random decoration. Once discovered and accepted by public awareness, an (temporary) intervention needs to transform and leave a mark on the site. Temporary uses represent an urban phenomenon that questions traditional planning or cultural heritage of a city[13]. It usually happens in places of uncertainties and schisms.

The "add on. 20 höhenmeter" project, initiated in Vienna in 2005[14] exemplifies such a metamorphosis of a site by an ephemeral transformation. The temporary structure converted during six weeks the Vienna Wallensteinplatz into a kaleidoscope of day-to-day life situations. It was next to a "vertical playground", an "artificial densification" and a "walk in environment" also "art in public space"[15]. The installation was an invitation to explore new relationships between private and public and its vertical component allowed users to discover new views of the surroundings and reinterpret the city. "add on. 20 höhenmeter" delivered

---

[13] „Jenseits des Plans, Zwischennutzung als Strategie einer neuen Stadtaneignung", Polis 2-3/2002

[14] http://www.add-on.at/cms/cat23.html. Accessed 02.02.2014

[15] Ferguson, Francesca. 2007. "Instant Urbanism – Tracing the Theories of the Situationists in Contemporary Architecture and Urbanism". Swiss Architecture Museum No 02: Christoph Merian Verlag

the illusion of a vertical compact city where people could experience an unexpected and mixed social environment.

*Fig. 6 "add on. 20 höhenmeter" project, Vienna.*

"The result was a fascinating diversity of perspectives of every-day life, of our surroundings in general, and more specifically of the location of the social sculpture itself"[16]

## 4. Synapse!

Once the site has been transformed it reveals its potential. It is vital to then enforce the connection to the rest of the city network. The site needs to find its added value beyond a temporary performance. Thus, it becomes a new link within the system of public flows, the synapse that expands the interconnected mesh of urban situations.

The best way of integrating the new site into the public realm is by programming it. Once programmed and engaged in the urban matrix, a more stable presence of the new (infra)structure can be insured. The function can be totally new in the case of urban voids without previous use or an overlapped program in the case of existing structures. The new program is aimed at refining the grain of mixed uses of the area and at increasing its diversity and attractiveness.

[16] "Instant Urbanism – Tracing the Theories of the Situationists in Contemporary Architecture and Urbanism". Swiss Architecture Museum (HG./ED.) No 02. Christph Merian Verlag

These uses can change over time and adapt to other needs. It is important for these sites to go on as activated joints of the urban structure and broadcast their energy towards new sites. One relevant and recent example is the one of the Ciclop park house. The first multi-storey long-term car park in Bucharest, Romania was built in 1923 and has ever since maintained its utility. Part of the representative street front, the building's facade is serving nowadays as an over-scaled publicity canvas, keeping its function in a discreet second plane. Its "crisis of identity" in recent years has opened Ciclop to the urban circuit. By becoming headquarters and selling point for a bicycle brand, the ground floor of the car park was revealed. The unusual activity draws attention towards the space beyond the façade. Nevertheless, it was not until the 2012 art in public space event "Street Delivery" that the whole space of the car park has been returned to the public. As part of the major street art event, the interior of the Ciclop building served as canvas for artistic manifestations of a selected group of Romanian artists. The parking areas turned into exhibition places and the big ramp, the backbone of the building, unfolds as a big exhibition strip for the art works. Turned into one big vertical art container and generator, the car park shifts its main use to meet the pedestrian public. Although many of the exhibited art pieces and installations have been removed after the closure of the "Street Delivery" event, the walls of the car ramps have kept the paintings. The car park and bicycle shop suddenly became an extension of the street. After the impelling event brought the site into public awareness, the imposing vertical space has been used as an alternative gallery as well. The three overlapped uses (car park, shop and gallery) are perfectly compatible and enrich considerably the quality of the serried public realm. The city space can be re-written by adding new layers to the spatial palimpsest. The Ciclop building will hopefully remain part of the street circuit as proof that vertical public spaces can be an alternative for alienating and disorienting city parts, where the built environment can engage with the protagonists of the urban experience.

*Fig. 7 The Ciclop car park during the "Street Delivery" event.*

## Conclusions

The purpose of the present investigation relies in justifying the need and finding a strategy for developing public spaces with a vertical component. Important influences from the 1960s, such as the enhancement of public awareness by the act of walking, the generation of situations that draw attention on possible sites of intervention, high flexibility for adaptive and changeable uses and the courage of overlapping structures and functions serve as ground for implementing the strategy of vertical public spaces. The important steps mentioned above can be resumed as follows: the discovery of the possible site, the activation of public awareness by the element of surprise and the creation of necessary urban connections by programming the vertical space. By that we can achieve a more sustainable urban environment with increased public activity, catalyse urbanity and densify uses within the urban fabric. Also, it is essential to discover the porosity of the surrounding structures in order to blur the sharp edges drawn between open public space and built environment. It is not about new buildings or wrapping up existent ones but about re- and over-programming existent ones. This strategy can work as a new engine for a more spontaneous and less radical generation of urbanity. It is vital to enhance the role of the street as a multi-functional urban space that accommodates a range of activities. The aim is to find an interface for these “deficient” components of the cityscape in order to resuscitate them into the network of public life within an ideal model of a city without standstill.

## References

Brook, Richard and Dunn, Nick. 2011. “Urban Maps. Instruments of Narrative and Interpretation in the City”. Manchester School of Architecture: Ashgate Publishing.

Careri, Francesco. 2002. “Walkscapes–Walking as an aesthetic practice”. Barcelona: Gustavo Gili.

Erickson, Jon. 1992.“The Spectacle of the Anti-Spectacle: Happenings and the Situationist International”. In Discourse, Vol. 14, No. 2, Performance Issue(s): Happening, Body, Spectacle, Virtual Reality, 36-58. Wayne State University Press.

Ferguson, Francesca. 2007. “Instant Urbanism – Tracing the Theories of the Situationists in Contemporary Architecture and Urbanism”. Swiss Architecture Museum No 02: Christoph Merian Verlag.

Friedman, Yona and Homiridis, Marianne. 2010. „Yona Friedman. Drawings & Models. Dessins & Maquettes“. 1945-2010. Les presses du réel, kamel mennour.

Gehl, Jan. 1996. “Life between buildings; using public space”. København: Arkitektens Forlag.

Gehl, Jan, and Lars Gemzøe. 2000. New city spaces. Copenhagen: Danish Architectural Press.

„Jenseits des Plans, Zwischennutzung als Strategie einer neuen Stadtaneignung“, Polis 2-3/2002 ,

http://www.urbancatalyst.net/downloads/polis_uc.pdf. Accessed 04.02. 2014 19:30.

“Urban Patterns for a Green Economy, Leveraging Density”. UN HABITAT for a better urban future.

Wallraff, Michael. 2011. “Vertical Public Space”. Nürnberg: Verlag für Moderne Kunst.

Iaac Closing Lecture: Yona Friedman Interview. 2011. Accessed 10.02.2014 10:00. http://www.iaac.net/lectures/maa-closing-lecture-yona-friedman- project ion-of-iaac-s-interview-197. Accessed 03.02. 2014 16:30.

Talks at the AA, AA Files 19 (Spring 1990), p. 32. Accessed 05.02.2014 14:30. http://www.cca.qc.ca/en/collection/283-cedric-price-fun-palace – Cedric Price. Accessed 03.02. 2014 13:25.

http://www.smarturb.org/index.php?#article:110, Time Based City. Accessed 07.02.2014 09:23.

# CHAPTER EIGHT

# CONTEMPORARY INTERVENTIONS IN HISTORIC PLACES: THE EXAMPLE OF THESSALONIKI METRO

STAVROS APOTSOS[1]

**Abstract** | The urban demands of each place are constantly changing as population, social and economic conditions alter. Cities, at different stages of their history, adapt to these changes through revitalizing interventions. Nevertheless, as technology, science and knowledge evolve, these interventions are performed in an increasingly greater depth. They become more decisive and more imposing, by revealing, altering or destroying the existing historical layers, or even by creating new ones often changing the urban landscape. The main issue of this paper is to examine the coexistence of contemporary design interventions with the preservation of historical identity.

In our effort to improve our cities, we face a wealth of parameters and factors that have to be taken into account. The cultural compatibility and the environmental sustainability occupy the highest positions in the hierarchy list of a vision for 2050. Sustainable development needs to include both the existing building stock and the urban infrastructure (roads & utilities) to support them. At this point, the present and future needs contradict with the past. Each intervention, either in building or in infrastructure, erases a piece of history. But in the case of the Thessaloniki Metro an unexpected discovery extended the route of the city on the map of history.

Excavations brought to light archaeological findings and impressive infrastructures of the ancient town; two roads were revealed exactly at

[1] Architect, Postgraduate student of Preservation and Restoration of Cultural Monuments, Aristotle University of Thessaloniki, Thessaloniki, Greece, stavros_apotsos@hotmail.com

their intersection. One of the central metro stations was designed on the eternal crossroads of the city, the "ancient heart" of Thessaloniki. The aim of this paper is to inquire into possibilities to preserve cultural heritage as well as enhance urban environments through elaborate schemes of conservation, maintenance and upgrade. In our case study, one step forward allowed us a thorough view backwards. And this gives us the opportunity to build our future on the knowledge of the historical past.

**Keywords** | infrastructures, contemporary interventions, restoration, urban history

## Introduction

Nowadays more than 50% of the global population lives and works in cities. Cities are the most imposing interventions of the human being on natural environment. The first metro system, "the London Underground", opened in 1863, just before its population exceeded 3 million people and the term "metropolis" was introduced to the world's urban literature. Today's larger city is Tokyo, with a population of about 32,450,000, while the second largest one, Seoul, is equipped with the world's longest metro system -"Seoul Metropolitan Subway"- with a total length of 733 km and 445 stations. Large engineering projects such as ports, airports, highways, subways and railways are the means that allow a city to cope with the massive demands of large populations. Habitation, employment, food and goods mass transfer, public transportations and all their utilities turn cities into engines of consumption, fulfillment of wishes and covering of needs. This is no different from what was happening in the first cities. The first cities were created near agricultural areas, where there was a surplus of goods. The human being lived, lives and will live in cities. Thus, every achievement of technology will always be embodied in them by any means necessary, in order to constantly improve the quality of human life.

## Interventions in Historic Places

> *The biggest difficulty in understanding a place is that our perception for that involves covered the concept of time. It consists of ruins, symbols, languages, experiential memories; it is never the same and so can not obey an easy typological analysis. (Giannakis et al, 2004)*

Man can conceive place as a book which maintains all human acts over time documented. This is also the reason that makes interventions in cities extremely difficult, while -when intervening- pieces of history are inevitably erased.

The identity of a place or a city is directly linked to the archaeological sites, monuments and historical totals. The preservation and protection of these areas in the centers of modern cities is therefore a necessity, but requires spatial and material sacrifices as well as complex manipulations, involving the association of these spaces with the modern city and the everyday life of the citizens and dictating ways of reading and viewing the symbolically charged Historic Places. Moreover, today, in a new system of globalization and "glocalization", the conservation, preservation and protection of these sites, as well as the search of differentiation or the specificity of each place acquire great significance.

The appearance of archaeological fragments and monuments in open public spaces is common in many centers of historic towns. In these cases the most important dilemma, which constituted the main design issues of the cities, was whether public space should be sacrificed in order to conserve and protect antiquities. The archaeological sites and heritage buildings are starting to leave their initial character - which wanted them to be isolated "sacred keepers" of the historical identity of the city – behind, and are integrated in evry day life. Through interventions that allow stopping, crossing, observation, perception, use and generally their experiential contact with residents or visitors, they lead to an organic integration with the environment and their surrounding urban fabric of the city in general.

The big bet that cities should win at the dawn of the third millennium - at a time when human communication does not necessarily require spatial contact - is to regain the realistic place, to rediscover their lost qualities, sensations and images in experiential memory, and all those elements which cannot be transferred by any electronic means, to finally rediscover the forgotten "morality of a place".

## Thessaloniki as a historic Place

Life in the Thessaloniki area dates back to prehistoric times. The oldest known findings date back to the Neolithic era and reach the beginning of the 3rd millennium BC. Since then, the human settlements around the Thermaikos gulf grew without interruption.

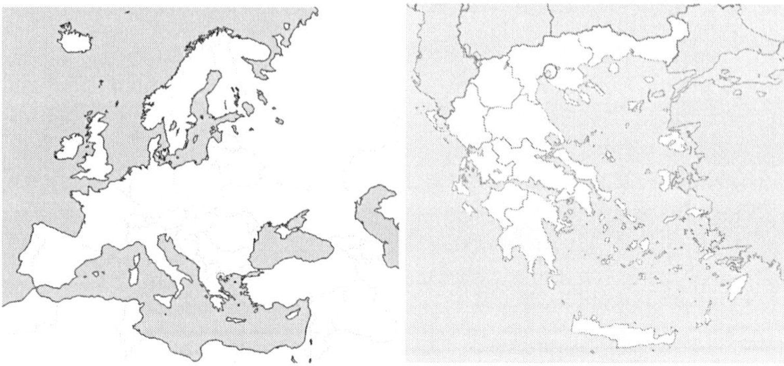

*Fig. 1: Location of Thessaloniki in Greece*

## History of the city

In 316 BC, just two years after he seized power, King Cassander of Macedonia founded the creek of Thermaikos Gulf, the city of Thessaloniki. To support the ideological position of the Macedonians, he named the city after his wife, who was half sister of Alexander the Great. In the 2nd half of the 2nd century BC, despite having conquered by the Romans, Thessaloniki goes on to become the capital of the Roman province of Macedonia and remains a thriving city for several centuries. In the early 4th century AD, Galerius, defines the city as the base of his territory and establishes his presence in the town (305-311 AD) with the palace complex and the known arch (Kamara), which is decorated with his campaign against the Persians. In 312 AD, Constantine the Great built an artificial harbour on the south side of the city as the most important, after Istanbul, economic and administrative centre of the Byzantine Empire, a role which the city kept until the domination of the Ottoman Empire in 1430. The economic and cultural development of the city was experienced only from the mid 19th century. Thessaloniki was liberated from Ottoman rule in 1912. In 1917 the historic centre of Thessaloniki was destroyed by fire. It was decided to be redesigned, work that was assigned to the French architect, Ernest Hébrard.

## Planning of Ernest Hebrard

The tactical planning which was followed by E.Hebrard used the monuments as reference points on the city's major streets. The main Byzantine monuments were arranged on the major axes, a fact that created an interesting urban landscape, but also minimized the possibilities of connection with their surroundings. The monuments and archaeological sites of Thessaloniki appeared to be submerged at lower levels under the growing modern city. This is a consequence of disasters and fires the city lived in the early 20th century, which destroyed the historic centre of the town. They were followed by large embankments, which altered the form and sank below the new level where the modern city was created and developed.

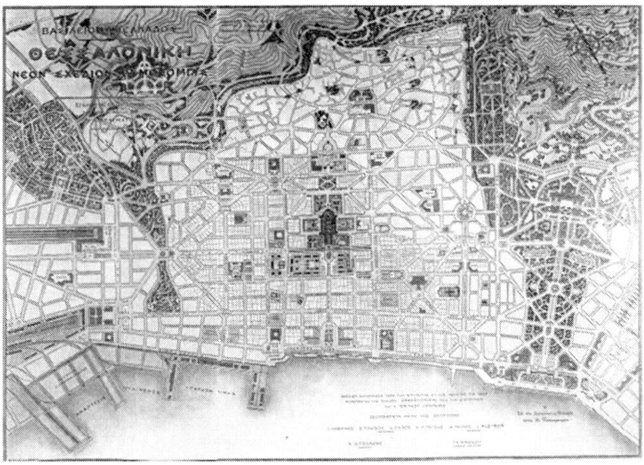


*Figure 2: E. Hebrard's masterplan of Thessaloniki*

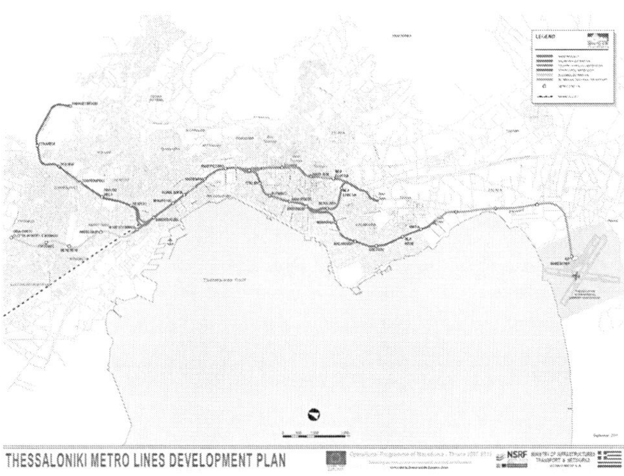


*Figure 3: Thessaloniki Metro Map*

Thessaloniki as a historic place consists of a plethora of monuments and archaeological sites. The most important of them were built in the Byzantine period. As in other cities, these monuments are fragments, which appear sparsely in the city and blend into the urban fabric. Especially, archaeological sites act as vital urban open space in an otherwise dense tissue. From this point arise two important issues. On the one hand, the modern city of Thessaloniki was rebuilt on the road axes assigned by the monuments. This design decision, taken by Hebrard was very important for maintaining urban typologies and qualities. On the other hand, archaeological sites were fenced and the monuments were excluded from public use. In an era (1921-1924), during which 97,025 refugees arrived in Thessaloniki, the demand for residential area expanded. This led to the sale of urban land lots and gradually to the development and reconstruction of the city centre. Within a short period of time, and in extremely harsh conditions, the debate for the conservation of monuments and archaeological sites seems utopian. There was neither time nor financial resource for such actions. Nevertheless, monuments and archaeological sites, although surrounded by a dense and tall urban fabric, survived. Nowadays, the construction of such a large project, such as the Thessaloniki Metro gives us once again, a unique opportunity to incorporate parts of the ancient city in everyday life.

## Thessaloniki Metro

The Thessaloniki metro is a revitalizing intervention, necessary for the sustainability of the city. The first phase of the project consists of 9.6 km of underground line (with twin tunnels) with 13 stations, and a depot at the east end of the line. Nowadays the project is under construction, and questions are constantly arising regarding the management of numerous archaeological findings of great historical value, which were brought to light by archaeological excavations.

Within over 7 years of archaeological excavations, more than 100,000 mobile archaeological findings have been revealed. Beyond these, the Golden Gate and the Vardar Gate, Necropolis, roads, cemeteries of the Late Roman period, Muslim graves and an early Christian basilica were also unveiled. However, the most important finding of all, which in addition has a great urban significance, was unveiled six meters below ground, in the place where one of the central stations of the Thessaloniki metro was designed, Venizelou station. From this point, the ancient Roman road "Decumanus Maximus" used to pass, which connected the east to the west gate of the city. At this point also, the "Decumanus maximus" intersected with a perpendicular street, "cardo", forming the commercial heart of the ancient city. Paved with marble, the road, that once hosted shops and workshops focusing on silversmithing, is still visibly etched by the ancient traces of passing carriages. Around the streets, there are building blocks that constitute the neighborhood of the ancient market. It is no coincidence that even today, the commercial activity on Egnatia Street pertains to silversmithing.

*Figure 4: Archaeological findings in Venizelou station*

Due to technical difficulties in the construction of the station, the company implementing the metro project, Attic Metro SA, proposed the transfer and, later on, the exhibition of the findings in another place.

## Conclusion

The construction of the Thessaloniki Metro revealed important information about the historical past of the city. Today, a unique opportunity is given to the city; to incorporate parts of the ancient city in contemporary, everyday life. The transfer of the archaeological findings to another place would be an irrevocable mistake. Not only because the architectural parts maintain their significance only in the place where they were designed and built, but also because the connection between the future and the past is a present issue.

## References

Giannakis, Ksenofon and Nikiforidis, Prodromos and Petridou, Kyriaki and Tarani, Paraskevi. 2004. *Thessaloniki upside- down.* Thessaloniki: Technical Chamber of Greece.

Global Health Observatory. 2014. "Urban population growth". Accessed January 23,2014.
http://www.who.int/gho/urban_health/situation_trends/urban_population_growth_text/en/

Lavvas, P. Yorgos. 1984. *Protection of Monuments - basic concepts, ideology and methodology.* Thessaloniki: Aristotle University of Thessaloniki - Service Publications

Ministry of Culture and Sports - Greece, 9th Ephorate of Byzantine Antiquities. 1997. *Modulations for 10 archaeological sites in Thessaloniki: The "unknown" City.* Thessaloniki

Organisation for the Cultural Capital of Europe Thessaloniki 1997. *Restructuring the City: International Urban Design Competitions for Thessaloniki.* Berkshire: Andreas Papadakis Publisher

Petridou, Zoi. 2005. *Archaeological Spaces, Monuments and Historic Ensembles in the centre of the city of Thessaloniki*

Proceedings of the 3rd National Congress. 2009. *Appropriate Interventions for the safeguarding of monuments and historical buildings.* Thessaloniki: Technical Chamber of Greece/ Department of Central Macedonia

SKG2020. 2010. "The solution lies in 1923!" http://skg2020.wordpress.com

Symbiosis. users.ntua.gr/kamy/symviosi.pdf

tovima. 2013. "Four proposals from Aristotle University for the antiquities of Thessaloniki Metro" http://www.tovima.gr/society/article/?aid=508533

tovima. 2013. "Thessaloniki: Over 100,000 archaeological findings in Metro" http://www.tovima.gr/culture/article/?aid=540497&h1=true#commentForm

Yerolympou, Karadimou Aleka. 1995. *The reconstruction of Thessaloniki after the Fire of 1917 - a landmark in the history of the city and the development of Greek urban planning.* Thessaloniki: University Studio Press – Publications of scientific books and journals

Yerolympou, Karadimou Aleka. 2007. *Thessaloniki, before and after Ernest Hébrard*

## Image Credits

Figure 1: *http://fenn-o-manic.deviantart.com/art/Europe-1925-Blank-Map-259329700*

Figure 2: *E. Hebrard's masterplan of Thessaloniki*

Figure 3*: http://www.ametro.gr/files/maps/AM_Thess_Metro_map_Oct14_gr.pdf*

Figure 4:
*http://www.grreporter.info/en/over_104000_finds_have_been_unearthed_during_construction_thessaloniki_underground/10304*

# CHAPTER NINE

# SPATIAL AND SOCIAL ASPECTS OF AN ARSENAL'S TRANSFORMATION: MILITARY PORT IN TIVAT INTO NAUTICAL-TOURISM SETTLEMENT AND PORT "PORTO MONTENEGRO"

GORAN RADOVIĆ[1]

**Abstract** | The sea coasts are very important places where one always tends to come. The coasts are the symbol of life, developing backbones, scenes and meeting places. Because of their importance, looking back in history, important strategic settlements were formed on the coastline with military, commercial and transportation ports. The ports gave those settlements significance and encouraged their development. Austrian naval port "Arsenal" in Tivat, built in 1889, was a significant overhaul port for more than a century. Naval port "Arsenal" in Tivat has made an important impact on the development of this small coastal place of Bay of Kotor. It has made an impact to demographic structure of the population, infrastructure construction, as well as the positive economic and social organization of the place. However, by its development and spatial spreading, the ''Arsenal'' occupied more and more space of this attractive coastal urban area. This area has interrupted a continuity urban settlement structure and it was a prohibited area, surrounded by 2 meters high wall. So, in a physical sense, it was an isolated urban area. By turning off of the naval overhaul company and by its selling to Canadian investor in 2007 begins a new period of development of this area. Namely, the area of the former ''Arsenal'', as a naval overhaul port, experienced a radical transformation that continued to this day, transforming this area into

[1] Professor PhD, dipl. ing. arch., Faculty of Architecture Podgorica, Bulevar Džordža Vašingtona bb, 81000 Podgorica, Crna Gora, goran.radovic@ac.me

attractive tourist complex called "Porto Montenegro" with luxury apartments, hotels, commercial buildings and becoming the most important aspect of nautical tourist harbour for mega yachts. This is an example of revitalization of captured urban area, which has been in the phase of abandonment and closure of military facilities in past few years, into the new development nautical and tourist base that opens towards the town and merges with it after a hundred years.

**Keywords** | Naval port, Arsenal, nautical port, transformation, Porto Montenegro.

## Boka Kotorska – history and cultural and historical heritage

On the southeast coast of the Adriatic Sea lies Kotor Bay, interesting because of its origin and appearance. The natural beauty and cultural heritage of Boka Kotorska Bay are somewhat sublime, unique and unusual and have inspired poets, painters, travelers, scientists and tourists. Due to its exceptional natural rarities and irreplaceable cultural and historical value, the Bay of Kotor is on UNESCO's list of world natural and cultural heritage sites and included in the club of the world's most beautiful bays. The wealth of cultural historical monuments of Boka Kotorska is evidence of the turbulent history of the region and it speaks of the presence of many civilizations and cultures in this region which have left behind significant architectural and artistic works of various epochal styles, dating from prehistoric times, through antiquity, early Christianity, Pre-Romanesque, Romanesque, Gothic, Renaissance and Baroque. The Bay of Kotor is drawn deep into the mainland between the winding slopes of the high mountain ranges of Orjen and Lovćen, forming several small bays, Herceg Novi, Tivat, Morin, Risan and Kotor Bay.

> *The overall harmony which is created by the bay's natural structure together with the coastal city's architecture was created from the timid respect of the people who have not dared to compete with the magnificence of nature. . . . The border between Mediterranean Europe and the mysterious Balkans has passed through here for centuries.*[2]

The shores are important places where man strives to come. These are areas that use geometric lines to separate land from water. The coastline is

[2] Sbutega Antun, Moja Itaka je jedan zaliv, From Arsenala to Porto Montenegro (Transformation), Monograph for the 11th International Architecture Exhibition - Venice Biennale Pavilion of Montenegro, p. 41, 2008. (My translation.)

a symbol of life, of the support of development, of scenery and meeting places.

> *On the banks, the water and land just touch each other; in the Bay, one element combines with another, the two permeating each other in the act of love. This interweaving of elements makes the bay become a new being, preserving the properties of land and water.*[3]

For these reasons, banks are strategically important areas where battles have been fought over the centuries and are still fought today. Because of their social and physical significance, important strategic areas with military, commercial and passenger ports were formed. Ports have given a distinct significance to those settlements and encouraged their development.

## Maritime Arsenal in Tivat – historical development

Such strategic importance and development incentives ware made by a military port, shipyard and arsenal in Tivat in the Boka Kotorska and Tivat Bays. The arsenal in Tivat was built as a military overhaul port for the Austro-Hungarian army in 1889. The development of the marine repair facilities of the arsenal resulted in the arrival of many diverse masters in Tivat. When a border between Italy and the Kingdom of Yugoslavia was determined, Trieste, Pula and Rijeka were assigned to Italy. A number of prominent masters came from these cities to Tivat Maritime Arsenal and there were a significant number of employees.

> *So, in 1921, the Navy Artisan School was opened and it turned into the Military Artisan School with a curriculum adjusted to the Arsenal's needs. . . . The two main production plants were machine building and shipbuilding clusters. In addition, there were maritime, chemical, economic and warehousing departments, and departments of weapons and ammunition, as well as a construction office. . . . Thanks to the planned training of its staff, the Arsenal was changing in both organization and production. . . . At the end of 1937, there were 1201 employees working for the Arsenal, 801 of whom were qualified, 180 semi-skilled, 133 unskilled and 87 serving apprenticeships.*

---

[3] Ibidem, p. 40. (My translation.)

*Fig. 1: Arsenal in 1918*

*For their results, the Arsenal workers were recognized by English experts in 1938, so the Arsenal gained a European reputation. It was possible to repair medium and smaller sized underwater and above-waterline boat parts, along with all auxiliary engines and machines in the Arsenal. In cooperation with the Artillery Technical Institute it was possible to perform minor repairs to all types of underwater weapons and to repair and overhaul artillery weapons of medium and small caliber. . . .*

*Fig. 2: Arsenal complex between the two world wars*

*Fig. 3: Arsenal facilities between the two world wars*

*At the beginning of 1941, there were 1562 workers, technicians, engineers, military officials, and non-commissioned officers and officers in the Naval Arsenal, and in the Military Vocational School there were three grades with 137 students. . . . After the Second World War the Arsenal was renamed the Naval-Technical Repair Bureau "Sava Kovačević" in which there were 1009 workers in 1960 and 1095 in 1970. In the sixties, the manufacturing plant for pontoon posts and anti-hail rockets was established.*[4]

*Fig. 4: Arsenal Panorama after the Second World War*

[4] The group of authors: Bojović Željko, Botica Ante, dr Pribilović Kažimir, Navy Technical Repair Bureau "Sava Kovačević" Tivat 1889-1989, Navy Technical Repair Bureau "Sava Kovačević" Tivat, p. 44-105, Tivat, 1989. (My translation.)

It was at this time that a two-meter-wall was built around the arsenal complex and the area was physically separated from the urban part. The high wall around the arsenal complex became an integral part of the urban vistas of the city. The wall "became an architectural event", in the words of Venturi, because it became "the border between the outside and inside". The wall is actually the controller of every activity in the area. The protective role of the wall is its primary role and control of the area was the reason for its construction around the arsenal complex, because the easiest way to control something is by the emplacement of physical barriers. For years men have fenced the areas they have wanted to control and in that way have dimensionally defined them.

In the seventies and eighties many non-aligned countries were sending their staff for professional training in the arsenal. The successful operation of the arsenal was reflected in the improvement of the standards of workers. So in 1974 a new restaurant with three buffets was built. During the eighties, the arsenal developed into a recognized and well-known repair center, experiencing a boom, and business and economic expansion. During this period, the repair facility within the complex invested in the construction of new public facilities, such as a garrison hospital, and the House of JNA with a cinema, library and two buildings for hotel-type accommodation of domestic and foreign crews during the overhaul of their ships.

In addition to investing in their own facilities within the complex, the Repair Bureau was actively involved in the development of cities and urban infrastructure. So it financially enabled, in 1987, the construction of a regional water supply from Kumbor to Tivat and the construction of a kindergarten in the village Seljanovo. The Repair Bureau provided support in the development of sport in the community. At that time, every other employee of the municipality of Tivat worked in the Repair Bureau, and based on this fact it was regarded as a business backbone for one hundred years. At the peak of its production and business glory, at the end of the nineteen-eighties, the Navy Technical Repair Bureau of Tivat was a recognized institution for repairing domestic and foreign military ships, repairing ships of the merchant navy, and production of war material and anti-hail rockets, and its specialty was the repair and maintenance of submarines for which it was recognized in Europe.

## Transforming the Arsenal complex into the nautical resort Porto Montenegro

Surrounded by a two-meter-high stone wall, the military repair Arsenal in Tivat occupies an area of approximately 30 ha of attractive inner city land.

> *Its construction and development, from 1889 to 2007, interrupted the basis of urban development of the city. Therefore, the development of the city mostly flowed in the hinterland of the coastal town and Arsenal. Forbidden to the public, the area of the Arsenal was the alternative zone in the development of the city, the untouchable island and barrier of its physical development. Shipbuilding related heavy industry prevented the development of tourism in this small coastal town and made the coastal areas and maritime zone ineligible for tourism.*[5]

On the other hand, it is evident that the institution of the Naval Technical Repair Bureau "Sava Kovačević" actively and substantially participated in the development of Tivat and its infrastructure and social facilities. Specialized military overhaul activity, over a period of a hundred years, affected the arrival and settlement of skilled personnel in the region and the formation and education of the domestic population in this activity, which greatly influenced decades of changes in the demographic structure of the population of Tivat. Besides, the good development and business status of the arsenal, which worked successfully for more than one hundred years, except during the war, positively influenced the social and economic population structure and significantly affected the overall culture of the lives of its residents. Such high levels of culture have remained present until today as a result of previous times, so even nowadays Tivat is known for year-round theater performances and concerts, the famous music and dance school, and many other cultural events. This inherited urban culture has resulted in the fact that, for example, today Tivat is the only municipality in Montenegro where utilities and municipal authorities have had no problems with the maintenance and billing of services.

Through the development of the military overhaul activities of the arsenal and its regional leadership position in this area, Tivat was a strategically important military port, which gave the city a military character during the last century. In 2007, when the complex of the Arsenal and Naval

---

[5] Radović Goran, Transformation: Arsenal–Porto Montenegro, From Arsenal to Porto Montenegro (Transformation), Monograph for the 11th International exhibition of architecture – Venice Biennale, Pavilion of Montenegro, p. 98, 2008

Technical Repair Bureau was sold to Canadian investors, the arsenal began a new phase in its history. The arsenal area, a former military ship repair brand, experienced a transformation. With the planned construction of the exclusive tourist complex "Porto Montenegro", with its mega-yacht marina, the arsenal area, finally, after 118 years, will become part of the urban part of the city of Tivat.

*Fig. 5: Demolition of the former arsenal*

> *The wall that separated the town and the arsenal was finally demolished. The isolated military area of the arsenal became blended with the city area. What was once a separated coastal belt in the city area would now become continuously associated with new residential and tourist areas with business and recreation facilities, which will ultimately valorize this area for tourism and give the city new high-quality facilities and contents.*[6]

The urban concept of Porto Montenegro as a tourist complex is planned as an integral part of the urban pattern of the city.

[6] Ibidem, p. 98, 2008

*Fig. 6: Boundary wall of the arsenal complex*

*Fig. 7: The scale model of the planned nautical tourism complex Porto Montenegro*

Designed as the biggest and the most modern Mediterranean marina for super yachts, with a superb range of services, Porto Montenegro has been designed in vernacular architecture, and claims to be the leading nautical destination in the world. Construction of the first part of the complex of Porto Montenegro has met expectations and the previously built complex has become an attractive and unavoidable place in Tivat. Besides, the construction of the nautical-tourist marina, with piers and docks, has attracted many foreign sailors and big spenders who anchor their boats here throughout the year. In summer, some significant persons come to Tivat and the Montenegrin coast. Construction of the first part of the complex of Porto Montenegro has significantly influenced the nautical and tourist appeal of Montenegro, which is now considered to be an attractive destination in the Mediterranean.

In addition to nautical and residential-tourist activities, in this modern transformation of the arsenal's space, some other activities and facilities were planned, such as business and commercial facilities, a museum, sports facilities and service activities as well as new infrastructure facilities. Construction of these facilities will complete the urban facilities needed in Tivat, and the arsenal area will be transformed into a new urban scheme included in the existing urban grid. New facilities in the former arsenal area, now Porto Montenegro, have already resulted in changes to the structure of employment, which again affects the demographic and social structure of the population. The transformation of the military overhaul facilities into nautical tourist facilities with the investment of a few hundred million Euros meant the transformation of the local economy too.

*Fig. 8: Newly built apartment facilities in Porto Montenegro*

*Fig. 9: Newly built Nautical-Marine facilities in Porto Montenegro*

The opening of new nautical tourism and business and residential facilities in Porto Montenegro has made Tivat an interesting destination with a lot of work places and well-paid jobs as well as a good place for professional training in this field and, overall, a good place for living.

> *The age-old history of the Arsenal and its unique spirit will be preserved through the nautical activity of the new marina for mega yachts and through the nautical museum which will be arranged in one of the old buildings of the former Arsenal. A warship and a yacht can express the symbolic transformation of the Arsenal, by war activity and hedonism.*[7]

*Fig. 10: Newly built apartment facilities in Porto Montenegro*

[7] Ibidem, p. 99

*Fig. 11: The newly built facilities of piers, moorings and promenade of Porto Montenegro*

## Conclusion

The Austro-Hungarian military port and shipyard of Tivat, built in 1889, was a significant overhaul port, which had an impact on the development of this small coastal place in Boka Kotorska and the demographic and social structure of the population for more than a century. However, with its development and the expansion of its capacity, the arsenal came to occupy more and more of the attractive coastal urban area, which disturbed the continuous urban development of the city and its tourism development, because the complex of the arsenal was placed in the town centre from the nineteen-seventies.

Closing the military overhaul enterprises and selling the arsenal complex to Canadian investors meant radical transformation of the former arsenal area and development will continue until 2020. The construction of nautical tourism and of commercial, residential and hotel and catering facilities have made this area an integral part of the urban structure of Tivat whose facilities are an open and integral part of the public life of the city. The demolition of the two-meter-high boundary wall around the former arsenal complex physically and symbolically presented the reassignment of this area to the urban area of the city.

In addition, the current construction of the Porto Montenegro complex has influenced the arrival of new workers and has resulted in demographic changes to this city that is growing from year to year. New facilities and a marina for mega yachts with their bindings, has attracted many nautical and rich tourists who come every year in increasing numbers, so the marine capacity is small in relation to demand. Tivat has newly experienced a transformation from being a site of military overhaul to become a nautical tourism destination visited by a growing number of well-known personalities during the summer season.

This is an example confirming that the transformation of space is an eternal phenomenon expressed through time and space as an inevitable event, using the language of architecture as a means of communication that physically expresses the transformation of the space.

## References

Grupa autora: Bojović Željko, Botica Ante, dr Pribilović Kažimir, *Mornaričko-tehnički remontni zavod „Sava Kovačević" Tivat 1889-1989*, Mornaričko-tehnički remontni zavod „Sava Kovačević" Tivat, Tivat, 1989

Kasalica Vladislav, *Kulturna baština Crne Gore*, Biblioteka Barski Ljetopis, Kulturni centar Bar, 2008.

Radović Goran, "Transformacija: Arsenal–Porto Montenegro", *Od Arsenala do Porto Montenegra (Transformacija)*, Monografija za 11. Intenacionalnu Izložbu arhitekture–Bijenale u Veneciji, Paviljon Crne Gore, 2008.

Sbutega Antun, "Moja Itaka je jedan zaliv", *Od Arsenala do Porto Montenegra (Transformacija)*, Monografija za 11. Intenacionalnu Izložbu arhitekture–Bijenale u Veneciji, Paviljon Crne Gore, 2008.

# CHAPTER TEN

# AN OVERVIEW OF URBAN REGENERATION PROJECTS IN TURKEY

MEHMET ÇETE[1] AND YUNUS KONBUL[2]

**Abstract** | Urban regeneration (UR) is an effective tool to make cities livable in the contexts of social, environmental and economic aspects. While some countries use this tool mainly to improve the economic conditions of cities, some others use it not only to provide economic but also social and environmental improvements to their cities. Turkey is a good example of the latter. Large illegal settlement areas emerged especially during the rural to urban migration period experienced after the 1950s on the outskirts of metropolitan cities in the country. Today, many of those areas remain in the city centers. Considering most of the population lives under earthquake risk in the country, the vitality and urgency of Turkish UR projects can be better understood. In this sense, a number of laws on UR have been enacted in the last decade. With the latest arrangements, Turkish authorities expect that UR projects will be carried out in a short period of time in many parts of the country. Currently, many local governments as well as the Ministry of Environment and Urbanization are beginning to develop and implement new projects some of which draw much criticism in different cities. Elimination of the issues experienced in those projects is of vital importance to the improvement of future projects. This paper, firstly, describes the legal bases and organizational structures of the UR projects carried out in Turkey. Then, it discusses the issues experienced in those projects. It is

[1] Assoc. Prof. Dr, Department of Geomatics Engineering, Faculty of Engineering and Architecture, Izmir Katip Celebi University, 35620 Balatcik, Cigli, Izmir, Turkey, mehmet.cete@ikc.edu.tr

[2] Research Assistant, Department of Geomatics Engineering, Faculty of Engineering and Architecture, Izmir Katip Celebi University, 35620 Balatcik, Cigli, Izmir, Turkey, yunus.konbul@ikc.edu.tr

concluded with proposals from a modern and academic point of view both to solve the issues and to improve current projects.

**Keywords** | Urban regeneration, slum, squatter, legal background, Turkey

## Introduction

Urban regeneration is a holistic approach to improving living spaces with the consideration not only of physical, but also social, economic and environmental aspects. Due to the insufficiency of physical and social infrastructure, economic problems, illegal settlement issues and so on, living spaces are becoming unhealthy and risky areas for communities. In order to renew these areas, authorities use various methods with different names. Today the most common and well-known method is called "Urban Regeneration" (UR) because of its comprehensive and holistic manner (Zheng, Shen, and Wang 2014; Couch, Sykes, and Börstinghaus 2011).

The main focus of UR projects was to improve the physical conditions of places but with less attention paid to the social side in the past (Andersen and Van Kempen 2003). However, in recent years it has become understood that focusing on the social side in UR projects is critical for their ultimate success. Otherwise, social unrest occurs; complaints and demonstrations arise (Uysal 2012). Today, such projects are started with consideration to ensuring community participation.

Turkey has been faced with a serious problem of illegal settlements (called *gecekondu* in Turkish), starting from the 1950s. Industrialization and mechanization in agriculture after World War II have caused rapid urbanization in the metropolitan cities (Baharoglu and Leitmann 1998). Agricultural employment shrank because of technological developments, which led peasants to leave their villages and move to metropolitan cities to find new livelihoods. The first comers built their shanties on the outskirts of the cities where land was low in value. However, the expansion of these areas was so rapid that the simple shanties turned into shanty towns in a short period of time. Even though earlier Turkish governments saw this occurrence as a threat to urban spaces and modern urban culture, no action had been taken. It had been ignored, even encouraged by many later governments and municipalities for political benefits, such as winning votes, or for humanitarian reasons as if *gecekondus* were a type of social housing (Baharoglu and Leitmann 1998). It had been welcomed by the private sector because squatters provided cheap and unorganized labor (Baharoglu and Leitmann 1998; Erman 2001). Several laws were enacted for the clearance of these areas. The first versions of the laws aimed to improve *gecekondus* to a good condition, to demolish others, and to prevent construction of new ones. The 1980s version of the law regarding *gecekondus*, known as "the amnesty law", actually went further from innocent provision of *gecekondus*, but instead led to their commercialization and eventually made them turn into four story apartment buildings (Erman 2001; Baharoglu and Leitmann 1998, Dündar 2001).

With the latest arrangements, Turkish authorities expect UR projects to be carried out in a short period of time in many parts of the country. Currently, many local governments as well as the Ministry of Environment and Urbanization are beginning to develop and implement new projects some of which draw much criticism in different cities. Elimination of the issues experienced in those projects is of vital importance to the improvement of future projects. This paper, firstly, describes the legal bases and organizational structures of the UR projects carried out in Turkey. Then, it discusses the issues experienced in those projects. It is concluded with proposals from a modern and academic point of view both to solve the issues and to improve current projects.

## Organizational and Legal Background of UR in Turkey

### Organizational background

There are several administrations carrying out UR projects in Turkey. They can be separated into two parts: local and national authorities. Local authorities are municipalities and special provincial administrations. The national authorities authorized for urban regeneration are the Ministry of Environment and Urbanization, and the Housing Development Administration (TOKI) (Uzun, Çete, and Palancıoğlu 2010). The Ministry possesses dominating power over any other administrations for any kind of development projects in Turkey. It can prepare, approve, modify, change or cancel development plans and projects of all kinds. The second national authority, TOKI, with the support of national government, has a high degree of authorization for urban regeneration (Figure 2-1). TOKI is today very active in the renewal of *gecekondu* areas and it has built around 88,236 apartment units within 14 years in UR projects. Statistics say that 85.93% of the projects carried out by TOKI are targeted to build houses for low and middle income people, 14% of which are UR projects (TOKİ 2014).

*Figure 2-1: An urban regeneration project carried out by TOKI (before and after) (TOKİ 2014).*

## Legal background

There are several laws relating to urban regeneration in Turkey. These laws are:

- The Municipality Law (03.07.2005);
- The Law concerning Renewal of Areas under Disaster Risk (16.05.2012);
- The Law on Renovation, Conservation and Use of Dilapidated Historical and Cultural Immovable Assets (16.06.2005);
- The Law on the Northern Ankara Entrance Urban Regeneration Project (04.03.2004);
- The Housing Development Law (02.03.1984); and
- The Development Law (03.05.1985)

According to Article 73 of the Municipality Law, municipalities are authorized to determine "urban regeneration project areas" to carry out UR projects in order to construct housing, industrial areas, commercial areas, public services, recreational and any type of social areas, and technological parks; renew outdated urban spaces; preserve historical and cultural assets; or take measures against earthquake risk. An urban regeneration area cannot be smaller than 5 hectares. Destruction, reconstruction or expropriation processes can be made with the stakeholders' mutual agreement. This may refer to participation at some point; however, the law does not mention any structural plan on the participation issue. Municipalities can expropriate any real estate in the project area, if necessary. Municipalities are also authorized to change or renew the facades of buildings, or to make environmental adjustments. Residents living in the project areas are subject to payment of only one-quarter of all kinds of development taxes.

The purpose of The Law concerning Renewal of Areas under Disaster Risk is the clearance and renewal of areas which are under the risk of natural disasters, or renewal of dangerous buildings. The Ministry of

Environment and Urbanization, relevant municipalities, special provincial administrations, and the Housing Development Administration (TOKI) are authorized to carry out renewal projects in such high-risk areas. The main focus is to renew dangerous buildings in a short period of time. The law seeks mutual agreement between administrations and owners of real estate that are subject to demolition. However, administrations can expropriate the real estate, if necessary. Projects that are carried out under this law are free from all kinds of development taxes.

The main purpose of The Law on Renovation, Conservation and Use of Dilapidated Historical and Cultural Immovable Assets is to protect, renovate, conserve and use dilapidated cultural and natural heritage assets in protection zones. With the authorization of this law, local administrations can construct housing, business, culture, tourism and social facilities; and take measures against natural disaster risks in protection zones. Mutual agreement is sought between authorities and persons for evacuation, demolition and expropriation. However, administrations can expropriate real estate in case of disagreement. This law is only applicable if the area is determined as a "protection zone". Protection zones are determined by the Cultural and Natural Heritage Conservation Boards consisting of scholars and experts. The projects that are carried out according to this law are free from all kinds of development taxes.

The Law on the Northern Ankara Entrance Urban Regeneration Project was enacted to carry out an urban regeneration project in a specific area in the northern part of Ankara and it authorizes the Metropolitan Municipality of Ankara to design and carry out the project. It is valid for a specified part of Ankara until the completion of the project. In other words, the law was enacted for a specific time, place and purpose.

The Housing Development Law authorizes the Housing Development Administration (TOKI) to carry out regeneration projects and to purify *gecekondu* areas. Normally, municipalities or provincial administrations are authorized to design and approve development plans; however, with the authorization of this law, TOKI can also design and approve development plans on its own in *gecekondu* areas with specific conditions. TOKI is also authorized with the right of expropriation.

The Development Law is the main law that provides fundamental principles and guidelines for development in Turkey. Municipalities and special provincial administrations are authorized to prepare urban and regional plans; regulate land use; and unite, separate, expropriate, sell and swop land parcels and ownership rights through Land Readjustment (LR).

In this law, there is no specific statement about urban regeneration; instead it is used as supporting material. Some scholars suggest that this law can also be used for urban regeneration. It is suggested that LR can be functional and cost efficient if the area to be regenerated has a low building density and if stakeholders volunteer for it (Turk and Korthals Altes 2010).

## Issues Experienced in Ur Projects in Turkey

Although the direct intervention of TOKI or its partnership with municipalities can produce faster and more effective results from the physical point of view, many TOKI interventions have received criticism for ignoring local communities' needs and causing gentrification (Güzey 2009). TOKI on the other hand defends itself by stressing that interventions are mainly focused on *gecekondu* and high-risk areas, where illegal occupancy problems and unhealthy living conditions are seen. In this respect, TOKI and other administrations responsible for UR projects in the country generally use the motto of "regeneration without gentrification" in their projects. The term is widely being used by politicians and mayors in order to stress that gentrification is not intended and that residents are encouraged to stay where they are.

The comments about UR projects carried out by municipalities vary as well. Municipalities, with the authorization of the Municipality Law and some other related laws, can determine "urban regeneration areas" with a decision from their municipal council, and start regeneration projects. Since there are different political agendas and socio-economic variations of the purviews of municipalities, the methods and results vary as well. The most common criticisms from the social perspective are lack of participation and gentrification. Increasing building heights and density are also criticized. Some UR projects were found very aggressive because of their gentrification-orientation and lack of participation (Uysal 2012), while some others were found rather successful because of the inclusive attitude of the relevant municipalities (Firidin Özgür 2013).

## Discussions

Any building, neighborhood or even district can be regenerated easily without any need for mandatory law or regulation if the stakeholders can agree on the terms. But mostly it seems impossible. The prioritisation of public interest, environment and public health can be provided by laws and regulations.

There is discussion as to why illegal settlers enjoy UR benefits; on the other hand "good citizens" who have not joined such illegal activities receive no such benefit other than what they have paid for. There is a fundamental philosophical issue here. Some people look at this issue as regarding state housing support to disadvantaged groups, because the Turkish State apparently failed to provide social housing in the 1960s to 1980s even though the Gecekondu Law envisaged providing social housing for low income communities. On the other hand, some other people oppose it and say that it rewards those who have got their land parcels by illegal means.

Are the laws in Turkey capable of enabling UR projects? There are many different laws addressing urban regeneration and they can be considered effective from the physical point of view. However they are lacking from a social perspective. They do not urge or enforce community participation. If all these laws are considered together, it can be understood that there are too many laws in Turkish legislation for urban regeneration. Although the Development Law is or should be the basis of any kind of development project, it seems to be pushed away by the regeneration laws. Güzey (2009) criticizes enacting split and partial legislation on development issues. She claims that every new law overruns the fundamental principles of the Development Law and damages holistic urban planning.

From the socio-economic point of view, there are very deep problems worldwide. There is a "downward spiral effect" in areas that are less favorable for many people and preferred only by those of low economic capability because of the low rents and prices offered in those areas. Therefore these areas are like centers of low income families. However, in a free economy, theoretically these people have the chance and equipment to break through their economic deficiencies. But when we observe the second and third generations of families who have grown up in the same neighborhoods, they actually show the same economic deficiencies most of the time. What lies behind this is a lack of education (Figure 4-1). It is common knowledge that people with low levels of education can generally only gain low incomes. While the needs increase together with the number of family members, economic deficiency converts into depression. In a depressive psychological state, people become more open to illegal activities, such as theft, robbery or other organized crime. When the media brings it to people's attention on the news, people from upper social classes stigmatize that neighborhood. Stigmatizing a neighborhood makes its residents lose their personal pride and sense of belonging and finally it converts into social unrest and radical activism. A child who grows up in such an environment develops a self-defeating idea that is called "learned helplessness" (Seligman 1975) which refers to a person's thinking that he/she is absolutely helpless and worthless whatever he/she does. It essentially destroys the desire of young minds to get educated. Individuals in this circumstance find it useless to have an education and that leads them to earn low incomes in their adult lives: and the downward spiral continues. Therefore, education opportunities should be increased persistently to break this negative circle. Authorities and non-profit organizations can provide educational activities for people. Suggesting increases in education opportunities is not a new thing; it was stressed, for example, at the 2007 meeting of the Leipzig Charter on Sustainable European Cities which consisted of the EU Ministers for Urban Development (Couch, Sykes, and Börstinghaus 2011).

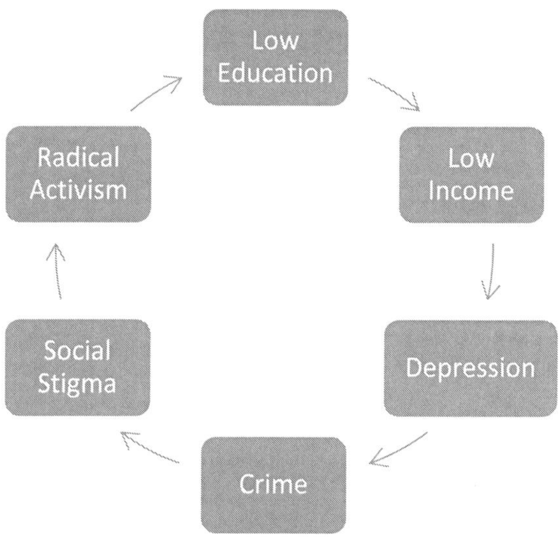


*Figure 4-1: Downward spiral of disadvantaged groups in deprived areas*

## Conclusion

Turkey has started the journey of urban regeneration with a top-down mentality just like other Western countries, but today has reached an understanding of the importance of using holistic methods with partnerships and participation. There are still many problems that need to be solved. National and local governments show their willingness to find better methods and repeatedly express their interest in public/private/

community partnerships and community participation. The public-private partnership method is being used more and more, because public capacity is not great enough for such large projects. The authors conclude that municipalities should act as a mediator between the communities and construction companies. However, satisfying every need is a difficult task and it takes time, so authorities and communities should be aware that the planning process may take a long time if needed. Planning processes can reveal better holistic solutions, so stakeholders should understand, encourage and advocate these processes rather than putting pressure on authorities to make quick decisions. Also, authorities should not put pressure on stakeholders to reach quick decisions either.

Helping out disadvantaged groups is a must according to current thinking and doing that can be possible through holistic urban regeneration projects that put people first to solve social problems by simply increasing job opportunities and providing social and educational activities. Authorities should try to overcome economic deprivation by implementing urban regeneration programs in these areas by simply generating spaces for shops, stores, markets or any other activities that can provide job opportunities. Also, education and training opportunities should be increased to break the negative cycle of "low education - low income - crime".

Community councils, boards and non-profit organizations have a positive impact on the processes and results of UR projects. The authors also conclude that these networks could be empowered with laws and regulations so that their positions and participation can be ensured.

Finally, the legal background of UR in Turkey seems to be mostly capable of carrying out UR projects effectively from a physical point of view. However the laws are very scattered. Having many laws on UR does not restrict the capability of the authorities but it causes confusion. Therefore collecting those in a single law, preferably in the Development Law, and supporting it with comprehensive regulations and guidelines might be a good idea.

## References

Andersen, Hans Thor, and Ronald Van Kempen. 2003. "New trends in urban policies in Europe: evidence from the Netherlands and Denmark." *Cities* no. 20 (2):77-86.

Baharoglu, Deniz, and Josef Leitmann. 1998. "Coping strategies for infrastructure: How Turkey's spontaneous settlements operate in the absence of formal rules." *Habitat International* no. 22 (2):115-135.

Couch, Chris, Olivier Sykes, and Wolfgang Börstinghaus. 2011. "Thirty years of urban regeneration in Britain, Germany and France: The importance of context and path dependency." *Progress in Planning* no. 75 (1):1-52.

Dündar, Özlem. 2001. "Models of urban transformation: Informal housing in Ankara." *Cities* no. 18 (6):391-401.

Erman, Tahire. 2001. "The politics of squatter (gecekondu) studies in Turkey: the changing representations of rural migrants in the academic discourse." *Urban Studies* no. 38 (7):983-1002.

Firidin Özgür, Ebru. 2013. "Urban design projects and the planning process: The Kadıköy Old Market Area Revitalization Project and the Kartal Industrial Area Regeneration Project." *Cities* no. 31:208-219.

Güzey, Özlem. 2009. "Urban regeneration and increased competitive power: Ankara in an era of globalization." *Cities* no. 26 (1):27-37.

Seligman, M. 1975. Helplessness: On depression, development, and death. A series of books in psychology: New York, NY: WH Freeman/Times Books/Henry Holt & Co.

TOKİ. *The Housing Development Administration* 2014. Cited February 6, 2014. Available from www.toki.gov.tr.

Turk, Sevkiye Sence, and Willem K Korthals Altes. 2010. "How suitable is LR for renewal of inner city areas? An analysis for Turkey." *Cities* no. 27 (5):326-336.

Uysal, Ülke Evrim. 2012. "An urban social movement challenging urban regeneration: The case of Sulukule, Istanbul." *Cities* no. 29 (1):12-22.

Uzun, Bayram, Mehmet Çete, and H Mustafa Palancıoğlu. 2010. "Legalizing and upgrading illegal settlements in Turkey." *Habitat International* no. 34 (2):204-209.

Zheng, Helen Wei, Geoffrey Qiping Shen, and Hao Wang. 2014. "A review of recent studies on sustainable urban renewal." *Habitat International* no. 41:272-279.

# PART 2:

# ARCHITECTURE AND TECHNOLOGIES

# CHAPTER ELEVEN

# INTEGRATED ARCHITECTURAL COMPLEXITY: FROM ABSTRACTION TO TECHNOLOGY AND MATERIALIZATION

RADA CAHTAREVIC[1], DZENANA BIJEDIC[2] AND AMRA TASO[3]

**Abstract** | Technological advances and new materials, information technology and new forms of communication, increasing urbanization and changing of social structures, along with increased complexity and transformability of architectural space, demand new conceptual models in planning and design. Today we have an environment formed by the combined influences of natural factors and technological evolution, lacking an optimal balance. The gaps between abstract thought and material dimensions, technology and natural processes, cultural memory and future possibilities, are not a recent problem of architecture. Integration of different systems, architectural with natural, should be achieved through the research and application of the principles of their complex dynamics. The complexity of architecture reveals analogies between architecture and nature at the level of complex systems' abstract principles. New concepts and new directions in the field of spatial planning and architectural design, through rejecting linear thinking by introducing a complex integrated approach at all levels, should produce a holistic result. The holistic concept of architecture considers space as a dynamic integrity of complex multileveled shapes, sizes and connections. New integrated conceptual models in architecture must be based on

---

[1] University of Sarajevo, Faculty of Architecture, Patriotske lige 30, radac@af.unsa.ba

[2] University of Sarajevo, Faculty of Architecture, Patriotske lige 30, dzenanab@af.unsa.ba

[3] University of Sarajevo, Faculty of Architecture, Patriotske lige 30, taso.amra@gmail.com

generative principles, integrating spatial planning, collective functions and global technologies with processes on the smaller scale, individual interest, the participation of local communities and consideration of local traditions. Synthesis of the top-down and bottom-up approaches is not some middle way, but must be comprehended as multi-dimensional integration, resulting in a complex dynamic balance.

**Keywords** | Complexity, holistic, abstraction, materialization, technology.

## Introduction

Environmental issues have evident impact on architectural practice, and although these issues have been discussed within architectural discourse, today's circumstances of the galloping progress of technology, rapid urbanization and the global connection of different cultures, create new circumstances and the need for a review of past practices. The ultimate goal of environmental advocacy tends to be to reach equilibrium within integrated social and natural systems, the built and unbuilt environment, past and future, and individual and collective needs, as a result of the complex dynamic balance of natural and man-made systems.

Technological advances and new materials, information technology and new forms of communication, increasing urbanization and changes in family structure, the conditioned development of new forms and functions, and interactions between and dependence on a large number of parameters, demand adaptability and transformability of architectural space. New spatial concepts generated in the field of architectural planning and design, with the aim of successful synthesis of architecture and nature, should transcend the formalism and linear thinking which have often placed architecture within closed borders, limiting its vitality, development and adaptability in a complex environment. These concepts should be established on the basis of research into both the theoretical and applied principles of these complex systems. The science of complexity could be seen as lying intermediately between the modern and postmodern views, integrating unification and plurality, theoretical generality and concrete cases of diversity and heterogeneity. Complexity science becomes a mediator between modernistic progress with its continual development of new and better technology, and the postmodern rejection of control and dominance over nature, placing more emphasis on sustainability than on linear progress (Emmeche, 1997).

## Complexity of Natural and Architectural Space

Although complexity is a ubiquitous and important theoretical concept, there is no general, unified definition of complex systems. Herbert Simon gives his definition of complex systems thus:

> *by a complex system I mean one made up of a large number of parts that interact in a non-simple way. In such systems, the whole is more than the sum of the parts, not in an ultimate, metaphysical sense, but in the important pragmatic sense that, given the properties of the parts and the laws of their interaction, it is not a trivial matter to infer the properties of the whole. (Simon, 1962)*

Complex systems consist of large numbers of individual entities with high interactivity that is mostly nonlinear, with feedback loops (Richardson et al., 2001). They are open to information flows both internally and externally. The border of the system is often difficult to define (Cilliers, 1998). Complex systems are flexible, adapting their behaviour to provide survival and efficiency, through learning and performing even more complex behaviours as a result of evolutionary processes. Among the properties that complex systems often possess are specifically self-organization and emergence. The overall behaviour of smaller components contributes to emergent patterns at a higher level.

Complex systems, although hardly predictable, can be modelled, and these models should not be based on properties of individual elements of the system, but on the complex relational structure of the system, revealing its dynamic patterns. The application of complex models leads to a shift from control and prediction to understanding.

> *Once we have a better understanding of the dynamics of complexity, we can start looking for the similarities and differences between different complex systems. (Cilliers, 1998)*

The study of natural systems is placed in a broader theoretical context in which the nature of such systems can be characterized, modelled and described by applying the theory of complex systems, and consequently, similarities and differences between natural and other complex systems can be defined (Parrot, 2000). Natural systems can be abstracted as a distribution network of related elements which operate by nonlinear interactions.

The theory of complex systems has flourished in the works of authors from the 1960s, but architectural interest intensified most greatly during the 90s. Knowledge in the field of complex natural systems can initiate the generation of new concepts in the field of architectural planning and design, with the aim of achieving a successful synthesis of architecture and nature. Studies of complex systems, especially natural systems that have perfected their design by adaptation through evolutionary processes, provide an inexhaustible source of new ideas that can be applied to architectural philosophy and engineering solutions. Considering architecture and nature as complex systems, presenting their complexity and interconnectedness on different levels, opens a new insight into the possibility of their harmonious integration as a holistic unity.

## Complexity and the holistic approach of traditional architecture

In the field of architectural practice, most contemporary architectural structures are increasingly based on a dependence upon advanced technological solutions. Such an approach generally leads to further extraction of non-renewable resources, and thus to an imbalance in the relationship between human and natural systems, complicating the processes of production and maintenance of architectural structures.

Due to the emergence of a large number of new interconnections in the totality of human life on Earth, answers to the problems of increased complexity are made possible by a holistic approach. "In the face of complexity, an in-principle reductionist may be at the same time a pragmatic holist" (Simon, 1962). Holism should not be here understood as a simplistic conception of the totality over consideration of individual components in some system that could be perceived as a sort of reductionism, but instead we must consider a more complex concept of holism. A holistic approach that could integrate totality and individualities must be based on an understanding of a system not only as the "composition of unity out of diversity, but also a composition of internal diversity out of unity" (Morin, 1992).

In the process of finding a perpetual compromise between the desired and possible, architecture is a field suitable for systematization and application of the theory of complex systems. The complexity of architecture, as a complex totality, reveals analogies between architecture and natural systems. Modern architecture has lost its connections with the natural environment, its processes, scales and rhythms. A holistic, integrated approach was mostly achieved in traditional architecture, which was based more on empirical experience, local cultural heritage and economic necessities. The morphogenesis of architecture was flexible, gradual and contextually conditioned by local natural resources. Traditional spatial patterns have been applied on many different spatial scales and on many functional levels. So their result cannot be seen as a formal composition but as a complex structure with many sub-levels and scales. Different levels of connectivity were not based on top-down broad plans but on bottom-up growth and the development of mutual relations based on cultural traditions, natural materials and the dynamics of everyday life.

Architecture in a post-modern information society discovers a new dimension through which it is possible to achieve integration with natural systems, at the level of the common characteristics of complex systems. Detailed analyses reveal that complex systems are completely different in their details and material substance, but at the abstract level they have important common features (Mitchell, 2009).

## Abstraction of space complexity

The concept of space is an abstract category, relying on perceptual, material and social aspects of human existence, while its scientific foundation is based on more rational geometric principles. Geometric space loses tangible physical properties, transforming itself into an ideal Platonic system belonging to the world of abstract ideas. The reliance of architecture on the abstract idealization of space, especially pronounced in modernism, often arises as a problem: the geometrization of space leads to a loss of touch with natural morphology and evolutionary processes, causing the separation of man from nature.

New spatial concepts following the theory of complex systems in the field of architecture, instead of the universal perfection of abstract rational forms, promote universal complexity as evolutionary transformability and incompletion (Cahtarevic, 2008). Geometry can abstract and reveal principles of the internal structures of complex natural materials and processes by the application of new fields of geometry, such as fractal geometry. The abstraction of complex systems enables the definition of unique principles that govern the dynamics and development of open systems, where all levels of development are correlated as the results of the laws of transformation that have multiplying and repetitive characteristics on different spatial and temporal levels.

For H. Simon, (1962) one of the basic properties of complex systems is hierarchical organization, which does not mean rigid formal hierarchy, but a much broader understanding of hierarchy. Different types of connections on different levels form a hierarchical network structure. Behind the very complex structures and forms of different systems, simple order can be revealed, transferred through many levels and scales. Complex abstract structures can be seen as a descriptive condition that is either the result of certain dynamic processes, or a generator of dynamic development: an algorithm that defines the dynamic development. Complexity emerges from non-linear dynamic development through several stages, where often one level of development becomes the starting point for transformation at the next complexity levels. Different levels of structural complexity could be dependent on initial conditions and on small changes through previous transformations. One primary state produces a multitude of different variations, allowing diversity from unique foundations.

The application of the abstract principles of complex systems, which in many ways reflect natural dynamic processes, enables the exploration and integration of the complex spatial dynamics of different systems. The integration of geometry with physical structures and materials, of abstract information with natural physical and biological processes, is possible only through a better understanding of the complex nature of their organization. Abstract complex dynamic structures can be used for simulation and analysis of complex processes, thus allowing analyses of different scenarios in architectural design, taking into account more complex processes than by linear planning.

Modern scientific achievements in the field of production of industrial materials, materials whose properties are the result of research into and application of abstracted natural and biological processes, show that abstraction as a principle of separation and generalization can be a reversed process (Weinstock, 2006). Generative rules governing the developmental dynamics of complex systems from an abstract level of spatial thinking could be applied to functional or material and structural aspects of architecture as the integral principles common to all aspects of an architectural structure's emergence as a complex form. This integration is based on the information-based computational equivalence of many dynamic processes, researched extensively by Wolfram, defining the four basic types of dynamic processes as homogeneous, periodic, complex and chaotic (Wolfram, 2002). Complex dynamic processes have the greatest creative potential, being open processes that could include many levels of different spatiotemporal scales. Complex processes have emergent properties which result from numerous individual actions at various scales of spatial and temporal development and maintenance.

## Integrated Architectural Design Process

In a complex dynamic system of the natural and designed environment it is not sufficient to optimize architectural subsystems and processes expecting that an environmentally acceptable built environment will occur by itself (Bijedic, 2012). Integrity of architectural design is not often achieved due to the absence of a holistic approach and the neglect of integral relations in the process of design. Those relations should be based not only on the linear correspondence between the form, function, material and the structure, but also on their complex interrelations and interactions with the environment.

The complex requirements of modern facilities result in truly complex programs at developmental space-time levels, including different sub-fields and sub-levels. Complex and multilevelled spatial, functional and structural problems in architecture demand synthesis of the multivariate dimensions, integrated into the overall spatial environment.

In traditional cultures, people mostly create their habitat by relying on natural instincts, providing solutions to many of the problems the world faces today (balance of eco-systems, waste recycling, use of local materials and construction techniques, human scale spaces, cultural identity, etc.). In this sense, a proper diagnosis of the current situation and understanding of the past should be a starting point for tracing the route of the future.

Other tendencies, which in certain ways are a continuation of human industrialization with its enthusiasm and hope for the endless possibilities of technology, have arisen in representatives of modern thought. The techno-centric approach identifies and indicates the presence of a problem of an endangered environment, and believes that the solution lies in continuous technological development.

> *Technology however is intrinsically optimistic: each new development, each new device brings the promise of a new future. Each new way of arranging atoms or bits opens the door to a new solution cloaked in a new form. And since these elements are infinitely configurable, technological development is more sustainable than sustainability, since it will never run out of ideas. (Fairs, 2013)*

## Nonlinear thinking in the development of technology and architectural design

The influence of man is so pronounced that today we do not have pure natural systems but hybrid integrated environments, nature-given and man-created, formed by the combined influences of natural factors and technological evolution. The integration of abstract thought with material dimensions, utility with aesthetics, technology with natural processes, cultural memory with future possibilities, is not a recent problem of architecture, which still continues to create an even wider gap between those aspects. It is the consequence of an unbalanced abstraction and virtualization of human life, a too rapid development of technology and an increasing dependence on information technology. Green movements may win some battles on environmental protection, but the future of human civilization cannot be preserved only by methods of saving energy and the use of less toxic materials. Technology on the other side is unable to solve problems by constant linear progress alone. “It would be a gross mistake to seek wholly within the field of technics for an answer to all problems that have been raised by technics” (Mumford, 1934).

By incorporating advanced technologies without moving away from the linear and quantitative approach to the treatment of architectural space and structure, architecture could become even more estranged from the natural environment, and more destructive to natural processes including human existence. A holistic approach that offers the possibility of perceiving architectural issues from multiple perspectives, in a far more complex way, can show us the way to respect the past, in which holism was a basic idea in creating architectural structures, while at the same time opening up new horizons to the future. The complexity of the dynamic process of the creation of architecture is becoming more pronounced, requiring constant evolution of ideas through their modification and improvement in the process of architectural design, demanding the joint efforts of heterogeneous teams of experts in various fields, ranging from spatial urban planning to energy efficiency. The process of architectural design rapidly changes, and at the same time generates new ideas, becoming a complex nonlinear process that was more intuitive in the past, but that today includes the application of technology, digital processes and the use of computer software.

Architectural building should never been seen as a hybrid collection of isolated fragments, but as a hierarchical spatial and material organization, in which the lower levels of the system are actively involved in improving and influencing the existence of higher levels of the system. Technology must be changed and integrated with natural processes on the basis of universal principles of complex dynamics. The involvement of specialists for different parts of architectural design and construction work should not be integrated only using commercial BIM software platforms. A more integrated conceptual approach is necessary to avoid further fragmentation on too many levels of an architectural project, and further dependence on software programming, an area not yet systematically explored regarding both its positive but also possible negative impact on architectural design.

Complexity science becomes a mediator between modernistic progress with its continual development of new and better technology, and the postmodern rejection of control and dominance over nature, with more emphasis on sustainability than on linear progress. Complex dynamics unfold as a continuous transformation leading to diversity and higher levels of complexity. Technologies and structural solutions could attain new possibilities in an integrated, complex holistic concept of architecture, considering space as a dynamic integrity of multileveled shapes, sizes and connections.

## Conclusions

Today we have an environment formed by the combined influences of natural factors and technological evolution, lacking an optimal balance. The aim of architecture is to achieve a balance within such integrated systems, as the result of a complex dynamic balance. Generating new concepts in the field of architectural planning and design with the aim of a successful synthesis of architecture and nature, rejecting formalism and linear ways of thinking, a new direction advocates the holistic integration of nature and technological society and their connections on multiple levels, by the improvement of design through complex evolutionary processes.

It is necessary to develop a new way of thinking, starting from education, research and social awareness, and only then could it be successfully applied in architectural practice. This new way of thinking should be non-linear, complex and integrated, regarding all spatial levels, abstract and material, personal and collective. New integrated conceptual models in architecture must be based on generative principles, integrating spatial planning, collective functions and global technologies, with processes on the smaller scale, individual interest, participation of local communities and consideration of local traditions. These principles should not be based on a linear way of thinking, so the synthesis of the top-down and bottom-up approaches is not some middle way, or on a common line of intersection, but it must be grounded on a multi-dimensional comprehension of their integration. Generating new concepts and new directions in the field of architectural planning and design, with the aim of the successful synthesis of architecture and nature, and rejecting linear ways of thinking, the improvement of design could be achieved through a complex integrated approach from abstract and functional to structural and technologically supported materialization at all levels of the architecture delivery process, resulting in a complex dynamic balance.

## References

Bijedić, Dženana. 2012. *Arhitektura, Holizam umjesto optimalizacije, Integralni pristup u arhitektonskom stvaralaštvu.* Sarajevo: Univerzitet u Sarajevu.

Cilliers, Paul. 1998. *Complexity and postmodernism: Understanding complex systems*. Routledge.

Čahtarević, Rada. 2008. "Universality of Complexity: From Geometric Spatial Concept of Modernism to Contemporary Architectural Form". *Prostor*: 16(1), 64–75. University of Zagreb

Emmeche, Claus. "Aspects of Complexity in Life and Science". *Philosophica* 59, 1997. 1 pp 41–68

Fairs, Marcus. 2013. "Sorry green design, it's over". *Dezeen magazine*, February 14th,
http://www.dezeen.com/2013/02/14/marcus-fairs-sustainability-technology/

Mitchell, Melanie. 2009. *Complexity, a guided tour*. Oxford University Press.

Morin, Edgar. 1992. "From the concept of system to the paradigm of complexity". *Journal of Social and Evolutionary Systems*, 15(4), 371–385.

Mumford, Lewis. 1934. *Technics and Civilisation*. Routledge.

Parrott Lael. 2000. "Incorporating Complexity in Ecosystem Modelling". *Complexity International*, vol. 07, Paper ID: lparro01,
http://www.complexity.org.au/ci/vol07/lparro01/

Richardson, Kurt, Paul Cilliers and Michael Lissack. 2001. "Complexity Science: A 'Gray' Science for the 'Stuff in Between'". *Emergence*, 3(2), 6–18.

Simon, Herbert. 1962. "The Architecture of Complexity". *Proceedings of the American Philosophical Society*, Vol. 106, No. 6. Dec. 12, 1962, pp. 467–482.

Wolfram, Stephen. 2002. *A New Kind of Science*. Wolfram Media, Champaign.

Weinstock, Michael. 2006. "Self-Organisation and Material Constructions", Techniques and Technologies in Morphogenetic Design. *Architectural Design*: 76 (2), 34–41. Wiley.

# CHAPTER TWELVE

# THE INTERPLAY OF MUSIC AND ARCHITECTURE: LAYERING OF SOUND AND SPACE

ANJA KOSTANJŠAK[1] AND MORANA PAP[2]

**Abstract** | A person can see, feel and touch architecture, but what if he / she could hear it too? Imagine an architectural music in which the surfaces, materials, and forms of a space speak directly, alluring the observer even further into affirmation of the spaciousness around them, evoking different memories and emotions. Innovative architects and designers across the world are finding inspiration in music to create sound-producing structures. These new, composite, living sound organisms, inspired by music, aim to heighten our senses and open our minds to the finer details that make up the world around us. As they become active, or alive, they reveal their own tones. Now through digital expression / digital technologies, architecture can attain new heights of creative supremacy. The aim of this paper is to explore, through interdisciplinary thinking, the connections between the aural and the architectural / the musical and the spatial, in order to achieve a better understanding of how space enhances our well-being. The question that arises is how sound / music influences the atmosphere of space, while at the same time showing that every environment has its own aural architecture.

**Keywords** | aural dimension, sound-producing structure, aural architecture, architectural music, digital technologies

---

[1] PhD student, University of Zagreb, Faculty of Architecture, Zagreb, Croatia, anja.kostanjsak@gmail.com

[2] PhD student, University of Zagreb, Faculty of Architecture, Zagreb, Croatia, morana.pap@arhitekt.hr

## Introduction

Close your eyes, open your ears. What do you hear? Silence or sound? If you are hearing something, then it is a sound. We hear it with our ears, it is a physical sensation and emotional experience.

Architecture evokes the inner meaning of space, through both a visual and an aural approach. It is important to take note that some architecture is made only to please the eye, neglecting the other senses that influence how we experience buildings spatially.

It has to be acknowledged that architecture can be sensed with more than just the eyes. Every architectural space has its own writing, it has the writing of all the senses, seeing, hearing, smelling and touching, (even tasting, in the case of the gingerbread house of Hansel and Gretel).

Examining the interplay of music and architecture, however, we inevitably depart from the visual and other senses that are intertwined with architecture. To enter deeper into the domain of music and architecture, we have to take a step backward, step away from music towards sound, and from architecture towards space. The question which arises is: "In which way does sound influence the sense of space, space itself, produce effects on the occupant of a space...?"

It is known that by hearing we can "see" the space around us. This space communicates with both architecture and music, bringing them closer to each other. While we walk through a space, with our eyes shut, we can still distinguish its parts. We are able to differentiate this space d according to whether it is full or void, big or small, open or closed. This "seeing", an ability believed to be vested in only some species of animals, is something of which we are all capable. We can all to some extent "feel" / "see" the space around us. For example we can divide our experience of sound in some space in layers - sonic perception, sonic event and passive acoustic object.[3]

In this paper it is our intention to interpret the interplay of music and architecture, through their intertwining relations, their spatial awareness,

[3] A sonic event represents an original sound, sonic perception represents an additional sound (the echo) and passive acoustic object can be for example a wall or floor or ceiling.

and their layering. They depend on each other; in a cave or in a cottage, expression is either melodic or rhythmic.[4] (Various, 2012)

## Disciplinary framework – considering sound as a material

In this section of the paper through a simple table a few common words and concepts relating to the sound will be briefly clarified. The reason for structuring / showing this in a table is that dictionary definitions of this words deviate from their acquired meanings and associations for this field research.

| | **Explanation of terms** |
|---|---|
| **aural** | human experience of a sonic process (parallels visual) |
| **hearing** | detection of sound |
| **listening** | active attention or reaction to the meaning, emotions, and symbolism contained within sound |
| **aural architecture** | properties of a space that can be experienced by listening (The major components of aural architecture include at least four types of spatiality: social, musical navigations, aesthetic, and symbolic.) |
| **aural architect** | someone who selects specific aural attributes of a space based on what is desirable in a particular cultural framework; acting as both artist and social engineer (focus on the way that listeners experience the space - cultural acoustics) |
| **acoustic architect** | a builder, engineer, or physical scientist who implements the aural attributes previously selected by an aural architect (focus on the way that the space changes the physical properties of sound waves - spatial acoustics) |
| **sound-installations** | sound-producing structures / sound interactive architectures / incorporeal instatement of sound within some restrained space |
| **sound-producing structures** | Composite, living, sound organisms, inspired by music |
| **sound-interactive architecture** | sound-producing structure / architecture as a result of the existence of sound-producing structure(s) |

*Table 2-1*

[4] A different development of sound treatment / music occupies different cultures, depending upon their living space. In cultures where people inhabited caves, the music they produced was more melodic than thatmade by people occupying pile dwellings, where the music was more rhythmic.

## Past researches: a brief history of the study of sound and space

Although we are inclined to believe that research into sound and space, the fascination and interest in them, started in more recent history, we ought to acknowledge that interest in both began a long time ago. The bond could be traced back to Vitruvius, who indicated the need for an architect to understand the harmonic properties of music from a mathematical perspective. But with respect to the bond of sound and space, we could go even further into ancient history, to mythical times, and mythical explanations of their relations, and the impressions they leave on humans and other beings with this kind of awareness, the influence that sound has according to the habitat occupied. (see footnote 4 in introduction)

Due to increased significance assigned to vision in the Renaissance era, with few exceptions[5] [6], architecture moved away from consideration of sound as an equal contributor to vision in its “creation”. Because of this turn of events, sound stayed remote, and during this period (from the Renaissance to the decades after the Second World War) was an outlying feature in the creation of architecture.

In more recent history, the relation between sound and space has fascinated many inquisitive persons and induced them to contribute through their research to the understanding of the bond and the interdependence between them. One of the main reasons for this increase of interest was the need for communication between architects and musicians, due to “the development of electronic sound transmission and production equipment, a focus on spaces for the performing arts within architectural culture, an increased tendency towards abstraction in musical compositions, the new omnipresence of the sound film as total environment, and the development within philosophy of phenomenology.” (Ripley, 2007, p. 4)

After the Second World War there were different approaches to sound and space. Some architectural spaces were designed entirely for the performance of music[7], while others were designed to bring better acoustic

[5] The Whispering Gallery, from the period of English Baroque, at the base of the dome in Christopher Wren's Cathedral.

[6] Athanasius Kircher first described the Aeolian harp or wind harp, an instrument known in ancient world

[7] The Phillips Pavilion at the 1958 Brussels World Fair, designed by Le Corbusier's office, with Iannis Xenakis as project manager, was made in order to house a multimedia spectacle

qualities to a performance space independently of the architectural space and vice versa. This could be commonly recognised as another version of total environment.[8] It is important to know that there is division of artists and thinkers who were concerned with the total environment outside of the realm of architecture, with a larger acoustic environment.[9]

The 1960s saw the formation of another field of artistic production, known as sound art. This field of art usually reacts to and interacts with the specifics of the observed space / given location.[10] On the other hand, some artists, like Bernhard Leitner, prefer the term sound architecture, referring to sound as a material used in the construction of space, in the construction of architectonic structures: walls, columns or domes made of sound.

## Recent developments

Nowadays it is important to acknowledge that the intertwining of space and sound has become even more attractive to young researchers across the globe. There are several different approaches to forming architecture / space with music / sound. Some approaches tend to form architecture through sound / sound installations, some build sound-producing structures that make sound from the forces found in the environment / from the forces of nature, such as wind or water, or even the sounds of the city / noise, while others build architecture sensitive to the aural approach.

### The flux of sound as material

We could say that Leitner was in a sense an initiator. He reigns in that realm on the verge of architecture, sculpture and music, allowing space to emerge through sound. He perceives sound as construction material, as a structural element of sound architecture, wall, column, beam, arc, dome and so on. Relying on the speed with which sound travels through space, on its continuous changes - back and forth, rising and falling, its dynamic

[8] A total environment could be explained through a theatre house, a building that must house all the senses and be capable of displaying them to viewers. A person shouting or whispering, walking on marble or on grass - every detail should be conveyed to the viewer, even felt through viewer as if he is experiencing it.

[9] It is closely related to acoustic ecology, and the study of urban soundscapes.

[10] David Rokeby's Very Nervous System, which uses body movements inside a room or restricted / constrained space to generate sounds in response to the dynamism of movements.

nature - he places it in an architectural framework.[11] Leitner speaks of "corporal" hearing, of hearing which is not only limited to our ears, but is felt and experienced through our body / our corpus, and as every part of our body is different, so is our experience of hearing.

Following Leitner's research, and many other sound explorations through various installations (sound-producing structures, sound sensible spaces) we are approaching sound sensitive architecture. Architects who bring sound sensitivity to their work.

## Architecture sensitive to sound

Renowned architects such as Steven Holl (Stretto House), Renzo Piano (IRCAM) and Peter Zumthor (Swiss Sound Box) have sought to incorporate ideas from the sonic realm into the realm of architecture. This approach however tends to be ignored in viewpoints of sound artists and sound architects.

For example, while designing Stretto house, Steven Holl did many sketches, but none of them felt right, so one day he asked a student attending the Julliard School of Music, what composition the default site remind him of, to which the student answered "Béla Bartók, Music for Strings, Percussion and Celesta, 1937. It is about overlap, about strata." With this new insight, Holl went to his office, and started designing through music, alternating between heavy and light, just as Bartók did.

| | Pictures | Name (of the project) | Author(s) | Location | Year | |
|---|---|---|---|---|---|---|
| 1. |  | **Chimes** | anonymous | Asia and Europe | ~3000 BC | ANCIENT H. |
| 2. |  | **Aeolian Harp (Wind Harp)** | anonymous, Athanasius Kircher | Ancient World,<br>Rome, Italy | ~1600 BC /~476 AD<br>~1650 | |
| 3. |  | **Philips Pavilion** | Le Corbusier office, Iannis Xenakis | Brussels, Belgium, EXPO '58 | 1958 | RECENT H |

[11] Through his researches into sound he has developed a sound cube, a restricted space formed through sound waves.

| | | | | | | |
|---|---|---|---|---|---|---|
| 4. | 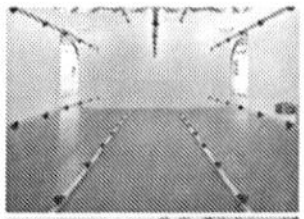 | **Sound Cube** | Bernhard Leitner | could be placed anywhere | 1969/ 70 | |
| 5. |  | **Stretto House** | Steven Holl | Dallas, Texas, USA | 1989/ 92 | |
| 6. |  | **Swiss Sound Box** | Peter Zumthor | Hanover, Germany, EXPO '00 | 2000 | |
| 7. |  | **Sea Organ** | Nikola Bašić | Zadar, Croatia | 2004/ 05 | |
| 8. |  | **The Singing Ringing Tree** | Tonkin Liu office Mike Tonkin. Anna Liu | Pennine mountains, Lancashire, United Kingdom | 2006 | CONTEMPORARY |
| 9. |  | **Aeolus Pavilion** | Luke Jerram | London, United Kingdom | 2009/ 11 | |
| 10. |  | **Organ of Corti** | Liminal, Frances Crow | London, United Kingdom | 2010/ 11 | |
| 11. |  | **Underwa-ter ORB Lights** | UrbanBotics | Copenhagen, Denmark | 2012 | |

Sound is an integral component of contemporary design, as these projects show.

*Table 4-1*

On the other hand, Peter Zumthor approaches sound and space from a different angle. Designing the Box he tried to create a total work of art, a space where you are meant to find placidity and also learn about Switzerland through dramatic music played by musicians moving through the pavilion, the food on offer, fashion and key words describing Switzerland "floating" in light, together merging with architecture.

All these approaches differ from one another, but are at the same time connected by the same appreciation for sound and space / music and architecture.

## City sounds

*"Wherever we are, what we hear is mostly noise. When we ignore it, it disturbs us. When we listen to it, we find it fascinating." John Cage. The Future of Music: Credo. Lecture, Seattle 1937. (Various, 2012, p. 45)*

## City sounds collecting

The interplay of music and architecture is best being approached through city sounds, mostly recognised as noise. The sound of the city is the sound of a car engine humming in the distance, a taxi horn cutting through the city smog, children's laughter coming from the kindergarten, the clatter of pots in a nearby kitchen, the roar of air handling units on the roof[12], etc. All these sounds represent life, in its fullest, but they also can increase the level of stress. Some of these noises tend to bother us more than others. Some people might even say that all these sounds combined are merely cacophony, and nothing more.

Questions that we should ask ourselves before dealing with refining[13] city sounds: "If there was a way to decrease city sound production / noise production, which means should we use and how?", "What might the city sound like in years to come?", "Can we use technology to retune the urban world that we now know?".

## City sounds refining

While thinking of refining the city's sounds we should also think of the impact we are making on the current soundscape, and what the outcome of our involvement might be. For example, we could use the experience of pubs after the smoking ban. When the smell of smoke 'left' those spaces, users encountered totally different smell in pubs. Taking a closer example for our focus on sound in the urban world, we could ask ourselves: "If the

---

[12] Sound artist Justin Bennett makes and collects audio recordings of cities, He does what could be called music, acoustic ecology, sound hunting, sociology, sound art, phonography or 'rhythm analysis'. (Various, 2012)

[13] Refining of sound as a mean of noise transformation. Transformation of noise into a more pleasing sound to the ear of a random listener.

sounds of traffic were refined, how would the birds sound?"" Would they sound even better, without the constant need to overpower the noise of the traffic?"

Another example of 'manipulating' / 'tuning the world' is the project Organ of Corti[14]. This project is in fact "an experimental instrument that recycles the sounds of our noise-saturated world, and is intended to challenge the definition of music." (Liminal, 2014, video) It allows users to become at one and the same time both composers and audience. On entering, you are engaged in a new way of interacting with the sonic world around you, and while standing there, a question arises in your mind, "What defines music as a piece?" This instrument "does not make any sound of its own, but rather it attempts to draw our attention to the sounds already present, by framing them in a new way". (Liminal, 2014, video)

Depending on the location in which such an instrument (or similar) is placed (by a lake, near a forest, on a city square, near a street or motorway), the sounds it transmits differ. The focus of our attention is on urban places and the impact of technology, which is becoming inevitable and more common today. Noises that can be caught and the new sounds that can be produced by such 'architecture' (sound producing or refining structures), can they change the wellbeing of the urban sound future?

## Conclusions

Aural architecture and auditory spatial awareness, when they are connected into one concept, provide a way to explore our aural connection to the spaces built by humans and to those provided us by nature. This fusion has lasted for centuries.

It has to be taken into consideration that it has been empirically and scientifically proven[15] that people do perceive space around them through the sense of hearing / auditory spatial awareness. It has been shown that people with temporary loss of hearing have difficulties in determining the depth of space. From this thought we can extrapolate that people sense the space around them through hearing, but this sensing is at a subconscious level, so that we tend to forget about the importance it has.

[14] The name is taken from the organ of hearing, which is located in the inner ear (organ of corti or spiral organ).

[15] In PLOS | ONE Lore Thaler from the University of Western Ontario, with Stephen Arnott and Melvyn Goodale, have reported brain imaging research that shows how individuals can echolocate, use what some blind activists call 'flash sonar', with which they accomplish perception neurologically. (Thaler, 2011)

With this knowledge, and many others, we are free to say that people do build space around them through sound, and with this paper, we open themes for discussion in the realm of sound and space / music and architecture.

> *"Listen. Can you hear it? The music. I can hear it everywhere. In the wind... in the air... in the light. It's all around us. All you have to do is open yourself up. All you have to do... is listen." Augus Rush (IMDB, 1990-2014)*

## References

Bernhard Leitner. 2014. "Work list." Work list. http://www.bernhardleitner.at/works

Blesser, Barry and Satler, Linda-Ruth. *Space Speak, Are You Listening? Experiencing Aural Architecture*. Cambridge: MIT Press, 2006.

Gresham College. 2007. "Music and Architecture: Confronting the Boundaries between Space and Sound." Lectures and Events. http://www.gresham.ac.uk/lectures-and-events/music-and-architecture-confronting-the-boundaries-between-space-and-sound

IMDb. 1990-2014. "August Rush (2007) Quotes." August Rush (2007). http://www.imdb.com/title/tt0426931/quotes

Inhabitat. 2014. "6 Examples of Innovative Architecture Inspired By Music." Architecture. http://inhabitat.com/6-examples-of-innovative-architecture-inspired-by-music/

Liminal. 2014. "Organ of Corti (2010-11)." Liminal. http://www.liminal.org.uk/organ-of-corti/

Mik, E. "Interview Steven Holl." ARCHIDEA, no. 42 (2010): 1-32.

Pallasmaa, Juhani. *The Eyes of the Skin, Architecture and the Senses*. Chichester: Wiley-Academy, a division of John Wiley & Sons Ltd., 2005.

Ripley, Colin; Polo, Marco and Wrigglesworth Arthur. "Introduction: In the Place of Sound." In the Place of Sound: Architecture | Music | Acoustics, edited by Ripley Colin, 1-14. Newcastle: Cambridge Scholars Publishing, 2007.

Thaler, Lore; Arnott, Stephen and Goodale, Melvyn. "Neural Correlates of Natural Human Echolocation in Early and Late Blind Echolocation Experts." PLOS | ONE, no. 5 (2011): 1-16.

Various Authors. *Music, Space and Architecture*. Amsterdam: Amsterdam Academy of Architecture, Architectura & Natura, 2012.

Zumthor, Peter. Misliti arhitekturu. Zagreb: AGM (Biblioteka Plan), 2003.

# CHAPTER THIRTEEN

# EVOLUTION DIGITALIZED: ARCHITECTURE OF SUBLIME DREAM

MIHAILO POPOVIC[1]
AND VLADIMIR MILENKOVIC[2]

**Abstract** | Although it might sound absurd, digitalization of the world is a natural process. Notions such as 'technological' or 'biological evolution', as well as 'digital' and 'post-digital age' have their origins in other disciplines. In architecture, the phenomenon of digitalization is clear and there is a straightforward methodological relationship between the real, the virtual and the actual, through all of which runs a spatial category. It is easy to see the broader context of the architectural discourse as cultural, political, economic or neo-liberal, super-modern or post-postmodern; but however the future of architecture is seen, it is crucial to redefine the spatiality of the human body, whose borders have been blurred by technology. In order not to step too far into anthropological territory, we will limit our exploration to concepts closer to architecture, such as the relationship between software and hardware, nature and technology, or myth and abstraction. States such as neoliberal capitalism or religiosity - or even more general ones such as a sense of abstraction or an inclination to the surreal - can be classified as natural ones, if one defines nature as an environment in which 'what survives is allowed'. Architecture, however, desires to redesign nature and, as such, always places itself in radical opposition to it. Architectural dreams always surpassed the possibilities of their fulfilment, thus growing into myths of power and infinity; we can see the transfer between the real and the virtual

[1] M.Arch., University of Belgrade Faculty of Architecture / Candidate at Strelka Institute, Moscow, Russia, popovic@edu.strelka.com
[2] PhD.Arch., Assistant Professor, University of Belgrade Faculty of Architecture, Belgrade, Serbia, link@arh.bg.ac.rs

that yields today's actual as a two way version of that same process, unmediated by other media. This means that it is Man's relationship towards himself which generates his relationship towards nature; and that Man understands his natural role by means of adjusting his dreams to himself. According to Heidegger, the word 'being' comes from the word 'to settle' or 'adjust' (the environment). According to Donald Mertzel and Meg Harris Williams, architecture is comprised of petrified dreams. Gilles Deleuze holds that the actual only exists when the real moves into the sphere of the virtual and gains a new purport. For Maurice Merlo-Ponty, invisibility is an integral part of perception, while for Juhani Pallasmaa it is more than that - it is a part of essence itself. According to Vincent Mosco, the sublime infinity is the necessary depth of the human image of the world, the same world in which the evolutions of the creator and the created, according to Bergson, became equal. Meanwhile, Decartes' infinite series of the unattainable returns us to the nature of Plato's primordial space.

**Keywords** | digital space, post-digital era, evolution, super-natural

## Semblance of Opposition

There is a notion that human intervention in nature is an act of opposition, yet at the same time an act which is essentially natural. In this mannerthe artificialandthe natural become conceptually polarized. The linguistic origin of the word 'artificial' it the word '*arte*'– art or by art, and '*factum*' – something created. From the beginning, the word had negative connotations: its accent is on something that is *not* natural, but has an intention to imitate nature. It is common to say of something naturally beautiful that it is beautiful as if artificial, and vice versa, to say of something artificial that is beautiful as if natural. Thus the polarities of the created (derived, manufactured) and the natural (made by the nature of things) are emphasized, especially when the former aspires to the latter. This relation is clearly visible in the process of translation in architecture, in the sense that the architecture is liberated from any mystification: the artificial is celebrated as an abstract category, or as a style in which the pure constructive endeavour of the industrial and post-industrial ages openly admires itself. Before, there were Corbusian airplanes, machines and ships – the ideas, concepts and overall aesthetics of the modern age; now, we have the endlessness of digital worlds and bioengineering that speeds up evolution; meanwhile, other things, like *the dream of flying* and its mythological capacities, still represent an idea of the sublime that is yet to happen. Returning to the relation that Man has with the object of his creation, we see that, etymologically, '*artefact*' takes us to creation, and the art of creation. These two notions are still elusive. According to Martin Heidegger, the very word *to be* originates from a word meaning *to create.* These words originated at a time when languages were still primitive, and meanings relational; at a time when ways of living – those created by man himself - were evolving: nomadic existence was giving way to a sedentary one based on agriculture, nature was being actively adjusted to create spaces for living and working. In this sense, construction in architecture is the basic anthropologic relation between man, his body and space. One should not forget that life on Earth started four times, although we have no evidence of previous evolutionary chains. As far as we know, the last skein of life to begin was the one of which we are part; the others remained written only in materials resistant to time, such as rocks.

## Dream, Myth and the Concept of Sublimity

For Man, the act of creation as such has always been controversial. Its sublimity was brought into question and Man faced the possibility and the fear of assuming the position of divinity: the audacity and responsibility for creating existence, the right and power to change the given and become a creator of new reality. The desire to overcome "pure" existence represents the infinite dream, by which the borders of reality can be overcome, that is, the circumstances resulting from the framework of existence. It is a desire to conquer the insuperable, to reach the heights, to leave a mark in time, to overcome the temporareity, not only of man as an individual, but of whole civilizations. In fact, it is about overcoming the dichotomy whereby the question of existence, duration, immortality and perfection is the central one. This aesthetic view is close to an understanding of architecture as *petrified dreams*. According to Donald Mertzel and Meg Harris Williams, architecture is a "solid dream", whose characteristics are "firm footprints" and a "full understanding of space", and whose goal is the absorption of time and the possibility of responding to one's sense of self. This relation with space in fact conceals a struggle with time. (eg. Daniel 24, *The Possibility of an Island*, Michel Houllebecq). Architecture, thus understood, is the construction of desire and belief – it is a sublime idea that, like a utopia, exceeds the possibilities of its realization.

> *"My wishes are grand, and expectations low." (Opera, Paris, Charles Garnier)*

Beliefs and myths carry with themselves a totalitarian component, and their danger is hidden in consequences that become foreseeable only when illusions fade, and myths transform into the great lessons (*The Way of Myth, Talking with Joseph Campbell* by Fraser Boa). Myth, on the other hand, also has a religious connotation: it involves belief in something that does not exist, or cannot be. Myths add a component of the virtual to reality, and are usually materialized as an act (ritual) and not an object. In the world of myth, too, architecture had always been a specific spatial-temporal object. Once grandiose objects of worship and divine temples were transformed, as civilization itself transformed, into the invisibility of financial systems - with their near divine characteristics - in which *god still lives in the form of money* (Giorgio Agamben). Financial bubbles are materialized mythological concepts, realized at the very moment when they are confronted with reality and burst. In this mythological sequence architecture is a medium of translation - neither goal nor topic. It is no accident that the last stock market myth was the one centred on real estate - the one in the not so distant year of 2008. The words that were supposed to symbolize lightness are actually physical objects of architectural weight, when removed from the idea of duration in its on-unfulfilled desires. (Babylonian tower) Contingency of Agambian type cannot be found in such concepts. The sublimity that would be a guarantee of aesthetic virtue in a given concept, even in the myth of money, entails separation of sublime subject and object. Therefore the idea of the eternal is always *outside* the subject, whose waiver of reality depends on uncertainty and the capacity for some other reality. The exact opposite holds, too: technology is just the transfer of imperfection, but also a simultaneous proof of human will.

> *"For it is no secret that under neoliberal capital, abstraction itself has become iconic, in a strictly religious sense. Nor can one be satisfied with opposing a resistant (or supposedly more real) architecture of brick to the apparently weightless balloons of financialization. For inside the empty vault, or in the real estate bubble or balloon, the air is thick and stale."(Reinhold Martin)*

For Reinhold Martin, the exit from this situation is engagement. For Vincent Mosko, meanwhile, the mythological character of and exaggerated belief in technology and digital infinity is illustrated by the financial bubbles that led to the crash of the stock market. This end(the bursting of the so-called ".com bubble")overlapped symbolically with the end of one century and the end of the second millennium. That means that the more the digital world and digital technology cease to exist as objects of our adoration, the greater their real power becomes. We are entering a period when, in a practical sense, the world of computers will become integrated in other things, and computers will disappear as objects in their own right, leaving worlds within the worlds, or forms within forms.

## Blurring of Body-Boundaries

Plato spoke of the human body's tendency to flee itself into the objects it creates. The transfer which occurs during creation symbolizes a narrow physical connection between the subject and object of creation. The linguistic logic of the title of Bergson's book *Creative Evolution* illustrates the blurring of the luminal values of the act we are trying to formalize here (L'*Évolution* créatrice vs. Evolusion creative). The overlapping of the subject and object of creation in a single image of evolution tells us that their relations are realized in technology, or in their manner of translation. Technology penetrates easily into our body, becoming part of it and degrading it, simultaneously erasing its borders of primary form: "heart beating in me today, can tomorrow beat in you, blood running in you yesterday, today runs in my body" (Stelarc). Mechanical hands that feel can be on opposite sides of the world. An artificial ear can hear where there are no natural ones. Due to digital technologies, eyes that see can access invisible spheres. Physically, we have become part of the digital world, and it has become part of us. The cyber and real world are blending into one. Some authors call this conception of the body a Cyborg – a cybernetic organism. This state redefines the relationship of mind and body. Stephen Hawking foresees the possibility of immortality by replacing the human body, probably with the form of some type of computer; while Michel Houllebecq still deals with issues surrounding the right to disappear. If we live in a time where our relationship to the human body is being redefined, then we must also be participants in the redefinition of the construction of its spatial frame. We consider *the other* construction, still seen in evolutionary terms as the adjustment of nature to human, while the structural role of architecture shifts from *meta* to *infra.* The concept of diminished visibility paradoxically enlarges its presence, and in renunciating the sublime we create conditions for the sublimity of renunciation. Along this line of understanding is also the utopian idea of Jeremy Till, according to which giving is made equal to freedom. If we put aside the purely philosophical implications of such an approach, we are left transferring architectural projections of the future into an image of future context, instead of an image of itself.

*Fig. 1 Perspective, M9 Master Project, Threshold of the Dream, Philharmonic, Natural Core of Belgrade, Mihailo Popović, UBFA 2013*

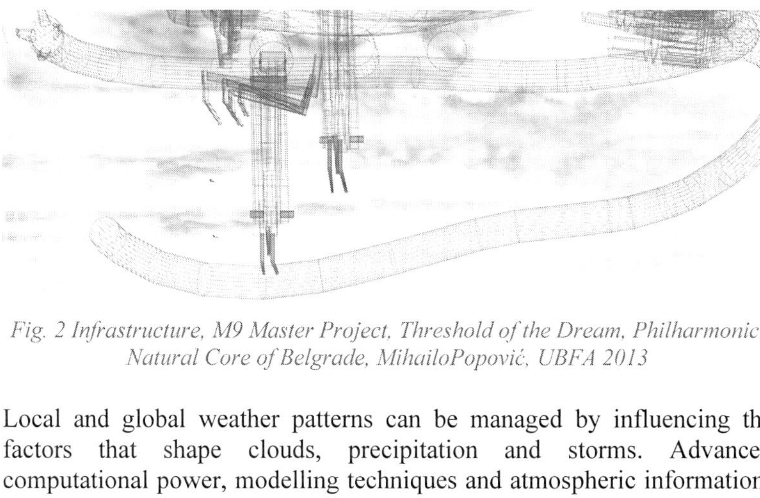

*Fig. 2 Infrastructure, M9 Master Project, Threshold of the Dream, Philharmonic, Natural Core of Belgrade, MihailoPopović, UBFA 2013*

Local and global weather patterns can be managed by influencing the factors that shape clouds, precipitation and storms. Advanced computational power, modelling techniques and atmospheric information-tracking will provide politicians with precise and systematic global weather modification capabilities for keeping control. These tools will be used to create optimum conditions for the human body – for its productivity and longevity. Should such tools fall into the wrong political hands, they could be used for destructive aims or military scenarios where people would have to dress up for the weather conditions. This system would mainly enable better food production in changing climactic conditions that promise bigger peaks and troughs, marked by floods and droughts. Agricultural genetic modifications in combination with climate control will provide local communities with sufficient food supplies, providing security for human bodies' basic needs.

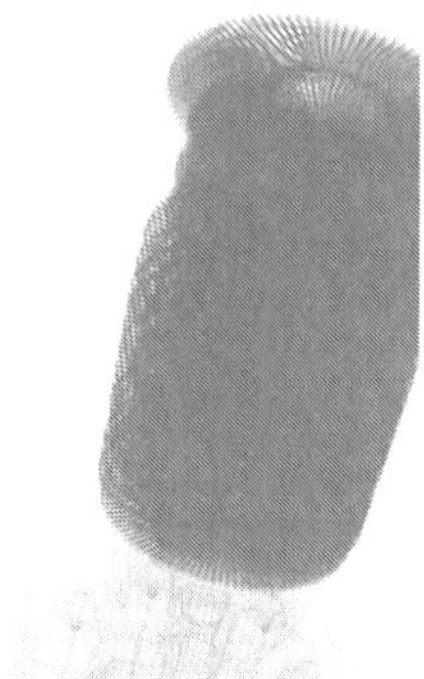

*Fig. 3 Fragment, M9 Master Project*

## Spiral Designing

The metaphor we utilised to signify the role of computers in architecture is built upon the feeling of infinity as the basic characteristic on which computer visions are designed. For design, the infinity of the Digital World is manifested as technology's capacity to translate the relative *uncertainty* of today into the relative *certainty* of the future. According to Mosco, cyberspace is mythologically established space, an area of crystallization of the natural feeling of being *between,* of appearing in the moment of transition from one state to the other. In that sense, analogy with the mythological sublime can "easily" become images and thoughts on the universe as a whole, magnified by the scale of these phenomena. If we accept that architecture is inseparable from its prevalence, we should accept that precisely that is its source of undoubted *naturalness*, and that transposing from a state that determines the other into *the other* occurs as a natural order of things. The naturalness on which we insist aims at placing the question of technology in architectural design conceptually within liminal areas of computer usage, in the zone where both the complexity and banality of this question are equal, in the place where it is possible to notice the real differences between the *being* of architecture and its role as a medium. It is still about the aspiration of the infinity of technology to rapidly change the image of the world. Besides virtualisation's mythological capacity, which is inherent to technology, for design it also means that abstracted reality has to constantly float above the virtual, so that universal values can stay beyond personal ones. Managing technology has never been a question of technology - even technologic concepts are created outside technology itself. According to Neil Leach, the contemporariness of architecture is a product of the direct connection between the manner of its presentation and its incidence. While this means that there is architectural content outside of architecture, it also points to the suspension of distance by which it is possible, in a traditional methodological sense, to control the presentation of its existence. In fact, we can question the whole view that it is possible to frame contemporariness, and that in the process of searching for its form it is possible to overcome it as a design problem with *the dialectic question of morphogenesis.* In this case, it is not the use of computers which is questioned, but the degree of naturalness of procedure. Leach considers that naturalness can be achieved by optimization, that that would have some kind of biological determination, and that the question of "beauty" would be covered in principle. The question remains, however, of how to generate meaningful coincidence using the repetitive mechanisms of machines, even when coding error is caused by the very procedure of generating new life.

## References

Bergson, Henri. L' Évolution créatrice. Paris: Les Presses universitaires de France, 1959.

Deleuze, Gilles, and Claire Parnet. Dialogues. Paris: Flammarion, 1996.

Heidegger, Martin. Poetry, Language, Thought. New York: Harper Colophon Books, 1971.

Houellebecq, Michel. La Possibilité d'une île. France: Fayard, 2005.

Martin, Reinhold. "In the Bank." *Thresholds* 41 (2013): 104-109. Journal of the MIT Department of Architecture. Accessed January 16, 2014.
http://thresholds.mit.edu/issue/41/t41.pdf

Meltzer, Donald, and Meg Harris Williams. The Apprehension of Beauty: The Role of Aesthetic Conflict in Development, Violence and Art. London: Harris Meltzer Trust, 2008.

Mosco, Vincent. Digital Sublime: Myth, Power and Cyberspace. MA: The MIT Press, 2004.

Paalasmaa, Juhani. The Eyes of the Skin. England: Wiley, 2005.

Savà, Peppe. "God didn't die, he was transformed into money." *Interview with Giorgio Agamben*(2012): libcom.org. Accessed January 28, 2014.
http://libcom.org/library/god-didnt-die-he-was-transformed-money-interview-giorgio-agamben-peppe-sav%C3%A0

Spiridonidis, Constantinos, and Maria Voyatzaki, ed., Teaching and Experimenting With Architectural Design: Advances in Technologies and Changing in Pedagogy, ENHSA-EAAE Transactions on Architectural Education No.35. Thessaloniki: Charis Ltd., 2007.

Stelarc. Art, Design, Future of Man. *youtube.com* (2012). Accessed January 16, 2014. http://www.youtube.com/watch?v=Y1SPish8ZwQ

# CHAPTER FOURTEEN

# BETWEEN THE PLACE AND NON-PLACE: ARCHITECTURE AND TERRITORY ON THE EXAMPLE OF SKOPJE

SAŠA TASIĆ[1], MITKO HADZI PULJA[2] AND MINAS BAKALČEV[3]

**Abstract** | It has been said that place is disappearing in modern contexts. The crisis of modern architecture first started with the crisis of specific places and the essential relation between an architectural object, technology and the object's place. But has contemporary architecture found its real place, or become even more disconnected from specific places? Has the contemporary conflict situation uncovered more dramatic relations between architecture, technology and places? The evident crisis of place in the modern era is only an anticipation of a permanent state of conflict in contemporary society. Certain theories already define the relations between historical places and super-modern non-places as well as between unique places and reproductions of them. But the main aim now is how to deal with the dynamic status of place in contemporary situations beyond the proposed binary relations of place and non-place and of original and copy. In the example of the city of Skopje, which went through a dramatic transformation during the process of modernization, it is possible to examine some aspects of contemporary places as seen through the issue of everyday places. Through selected projects we will examine possible approaches to places and technologies, which go beyond

[1] MSc Assistant Professor University Ss. Cyril and Methodius, Faculty of Architecture, Skopje, sasatasic@gmail.com

[2] Dr. Professor, University Ss. Cyril and Methodius, Faculty of Architecture, Skopje, mitkohadzipulja@gmail.com

[3] Dr. Professor, University Ss. Cyril and Methodius, Faculty of Architecture, Skopje, minasbakalcev@gmail.com

the dialectic of place–non-place towards the notion of the territory as principal medium of architectural modification.

**Keywords** | place, non-place, everyday place, territory

## Introduction: Place and Non-place

The contemporary crisis of place is a result of a fundamental change in the positioning of the architectural object and its location. A number of authors have examined the correlation of architecture and context in the modern world from different points of view: as life process and physical structure, geometrical space and anthropological space, as a utopia and heterotopia, as unique place and its reproduction. In the 1990s Marc Augé wrote about a changed perception of place in the time of super modernity, while Walter Benjamin in the 1930s wrote about a change in the relationship between an object and its place in the period of mechanical reproduction (Augé 1995; Benjamin 1968). Marc Augé made a distinction between anthropological spaces and modern places, between place and non-place. They set the opposing poles of the super-modernism of today. But this type of approach opens up a range of issues in architecture.

According to Mark Augé, place has at least three common characteristics: place of identity, relations and history (Augé 1995, 52). A place is a constituent of individual identity. A place of any kind is characterized by containing an order under which the elements are distributed in relation to coexistence. A place is historical in the sense that history is part of the collective and individual practice.

Opposite the spatial constructs labelled as places are non-places.

> *If a place can be defined as relational, historical and connected with identity, then a space which cannot be defined as a relation, or historical, or connected with identity will be a non-place. (Augé 1995, 77-78)*

The hypothesis that Mark Augé sets out says that super-modernism produces non-places, meaning spaces which themselves are not anthropological places and which fail to integrate former places. If in this paradigm, places are closed pre-modern ethnic cultures, indigenous societies, local technologies, precisely located in space and time, then non-places are an expression of individual personality, applicability, temporality, global technology and production: ephemeral transit points in which "luxury and inhumane" conditions proliferate, a world of anonymous airports, hotels, supermarkets and informal configurations.

Recognizing the contradiction of place and non-place, Mark Augé indicates their ambivalent relationship: that neither of them exists in a pure form. That ambivalence is the basis of this article, which is aimed at exploring the area between and also the positioning of the place in relation to the technology that produces and reproduces it, an area which is certainly recognized but has not been sufficiently explored and explained. The purpose of this research is to position an everyday place as resistant to contemporary practice contrary to imposed of binary discourse of the place. The assumption is that in today's marginalized everyday places, it's possible to have reconstruction and recycling of a sequence of inclusive tactics and technologies as a complex and eclectic system of relations, subsequently differentiated through layers of reference of the site.

## Dual Places

In the nineteen-thirties, Salvador Dali got a postcard that shows an African village. When he rotated this postcard by 90º, in it he saw a phantom head, a Portrait of Picasso. That association, to see something otherwise or passing from one state to another, transforming to another representation, or a prepositional start causing a series of associative arrays, basically presented the paranoid-critical method which Salvador Dali formulated in the nineteen-thirties. He produced challenging pictures with surprising and authentic natures by transcending the validity of reality (Dali 1936). This method was a result of the technique of reproducing one authentic place in a multitude of copies and in a surprising way, introducing us to the world of continuous flux of the authentic associations of unique images, in which from one picture another emerges, from one state another is created.

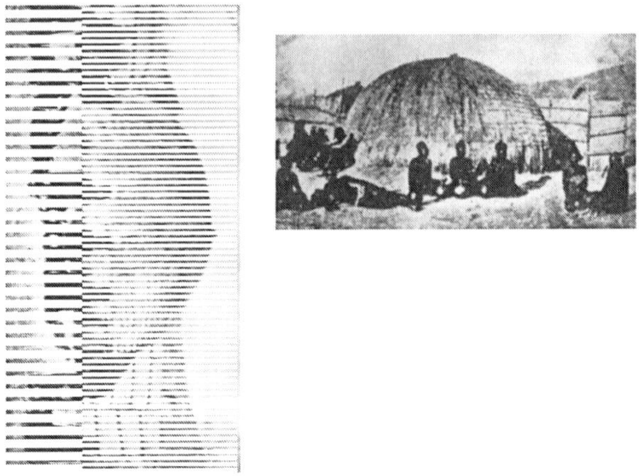

***Figure 15-1.*** *Dual Image, two positions of the postcard of an African village, which in the early 1930s was received from Salvador Dali by Pablo Picasso. The opportunity to see two things in one picture, a village and a person, served to illustrate to Salvador Dali the concept of the paranoid-critical method (Rosenberg 2012).*

In a similar way we can follow progressive visions of modernity. Le Corbusier's book, *The City of Tomorrow*, gave an evocative vision of the new city as a juxtaposition of a setting of one spatial order and the context of existing capital (Le Corbusier 1971, 276-289). In the plan "Voisin" 1925, in one frame two pictures are shown, the district that should be demolished and what is suggested to be built in its place. In a dramatic way it presents a juxtaposition of the simultaneous presence of the imaginary versus the existing. Actually the two pictures show that this is an extreme scenario for the place, which can be seen in two ways, and they refer to two opposite realities, one that is historically rooted in place and another that transcends the historical model of the site and proposes a new spatial structure for the city of tomorrow. The two pictures are supported by two different technological paradigms: The traditional paradigm that develops architecture in a chain of time sequences, and the modern paradigm, where architecture is based on an instant version of the serial production of one element. The possibility of the appearance of one place multiple times in vertical placements – as in skyscrapers of the 1960s, the new architectural typology suggested by Le Corbusier – problematized the authenticity of a place in a dramatic way.

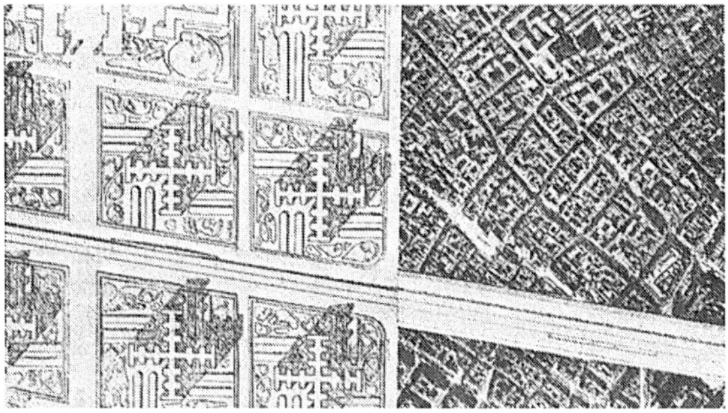

*Figure 15-2. "Here is the solution proposed by the 'Voisin' Scheme. Here are the districts which it is proposed to demolish and those which it is suggested should be built in their place. Both plans are to the same scale." (Le Corbusier 1971, 289).*

In this way modernity has questioned the historical relationship of architecture and place: in spatial plans through revitalization of specific spatial situations and their places, and in technological plans through the alignment of the methods of building and technology over particular places. However the real result of the project of modernity is not only to repeal the existing "backwardness" of the physical structure and impose a new set of opposing spatial lines; the net result is precisely the fragmentation of existing spatial situations and the introduction of the simultaneous presence of different fragments, of different spatial models and technologies of building, as possible configurations of a place. In this way the fragmentation of our cities cannot be understood as being a result of the exterior appearance of modernism's architectural examples, but as a result of the essential re-conceptualisation of the place through the relationship of the architectural object and the technology that produces it.

## Everyday Places - Recycling Technologies

In the example of the city of Skopje, which went through a dramatic transformation during the process of modernization, it's possible to examine some aspects of contemporary places as seen through the issue of territory. Through selected projects, we will examine possible approaches to places and technology which go beyond the dialectic of place–non-place, towards the notion of the territory as the principal medium of architectural modification.

What do we see today in the city? Today's city is all but a continuous physical artefact. A view of Skopje (2x2km) shows heterogeneity and a divergent texture: almost tactile, we can feel the differences. What is behind this un-homogeneous picture? The answer to this question was the real reason for the research into the city's morphology, through a series of analytical drawings. In this manner, within one frame we have decomposed an array of thematic layers.

***Figure 15-3.*** *Skopje, city of fragments/collection of different pieces*

In this way the city in the given frame is exploded into a city of fragments, the product of a series of reflections, a series of exclusive lines that once had to be established and form and reform the city, which as a result produce an incoherent picture of its texture (Bakalchev 2004). It appears as a city made up of cities, a city of possible worlds.

***Figure 15-4.*** *Novo Maalo neighbourhood, on the poster Architecture of Everyday Life, 16 Biennale of Macedonian Architecture 2012; Badnik (Christmas Eve) fires in the Skopje neighbourhood.*

However, if the view from above shows the complexity and contradiction of the modern city, what is the view from below, the view of an urban stroller in his everyday places?

The poster of the Biennale of Macedonian Architecture from 2012, titled "Architecture of Everyday Life", shows a picture of one of Skopje's neighbourhoods. Specifically it is a picture of the Novo Maalo neighbourhood, one of the pieces of the city, which almost consumes itself, dissipates before our eyes. What's in this image? An ensemble of once active houses, the remains of one nostalgic past: but what truly possesses a deep human scale is the humanity, regardless of the complexity, bizarreness, pretence and the poverty by all appearances. Although the neighbourhood evidences devastating processes from the lengthy transition process of the urban fragments of the city, it still has a spatial and social capacity; it has the specific quality and atmosphere of an authentic place – historical, relational and identifiable. On Christmas Eve, in different neighbourhoods bonfires are lit and residents gather around them. This one tradition, although increasingly subject to criticism, shows the authentic desire of the residents to belong and connect to the place. In Novo Maalo, Christmas Eve is celebrated at a specific place, the intersection of five streets, whose central axis touches upon the modest neighbourhood tap, while leaving "contemplative" spatial pockets. This place genuinely holds authentic value, on the one hand common and everyday and on the other archetypal and timeless. Passing through it, it feels as though a great star appears under our feet with the power to organize and at the same time transcend our daily spatial experience.

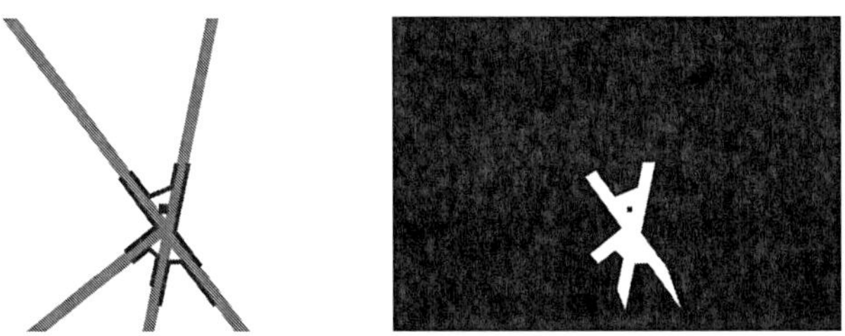

***Figure 15-5.*** *"Star": intersection of streets in Novo Maalo*

This neighbourhood, Novo Maalo, was the subject of research by the Integrative Studio of Architecture of University Ss. Cyril and Methodius. The original project of the architectural studio, "Tactics of transforming residential texture: a new collective form of Novo Maalo", was followed by individual Master's studies aimed at exploring ways to transform an urban fragment or residential gap, deriving both from everyday hypothetical situations and the visions of utopian energy of the seventies through Fumihiko Maki's concept of collective form (Maki 1964).

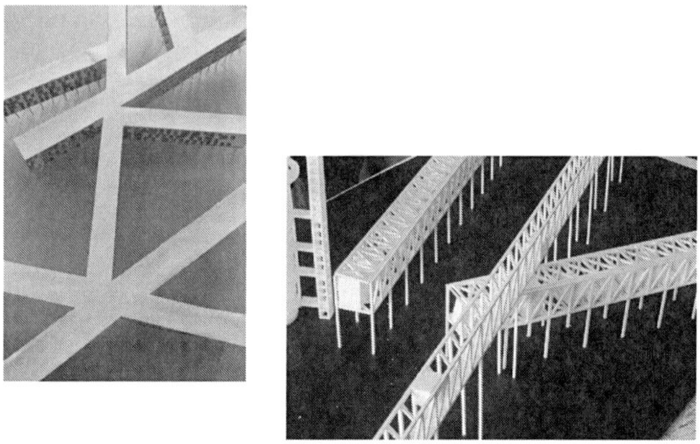

***Figure 15-6***. *Models of integrative studio tactics for transformation of the housing texture, a new collective form in Novo Maalo neighbourhood, Skopje, University Ss. Cyril and Methodius, 2013*

In the project we started from an inversion of the spatial neighbourhood system, kind of an anti-neighbourhood. What if the void of the streets becomes solid? What if the solid texture starts to become void? The series of projects have given different alternations on this inverse reading of the neighbourhood: raising of the streets, layering of the streets, upgrading of the streets, sequential connecting of selected land plots and their extruding parts, cutting of the city as a method for interaction of the urban fragments. All these projects were supported by technological models developed in modern times, standard structural systems based on repetitive modules, and the use of infrastructure structural models, now placed in closer local situations and contexts. In this way, the technological models are freed from an ideological basis on one particular hierarchy, from above to below, facing the local context with a new and productive intensity.

All of them proved to be a dialogue between the historical place and the new place, a kind of double dynamic place, a new structure layered on an existing one, but not excluding the existing, rather including, upgrading and assimilating. Tactics using and recycling existing technologies brought back a dialogue with the place. These same tactics bring us back to the primordial character of a place as a territory in which different configurations are only inclusive episodes from which exciting new spatial constructs can become.

***Figure 15-7.*** *Aleksandar Petanovski: linear configuration as incision, cutting the city (MA Project), University Ss. Cyril and Methodius, 2013*

## Conclusion

To go back to Salvador Dali's paranoid-critical method, in part it was based on challenging with images of a surprising and authentic nature by transcending the validity of reality. This method leads to a world of continuous flux of associations; from one picture another emerges, from one state another is created. Similarly, looking at the city today inspires a new critical interpretation of reality. Salvador Dali trans-edited the technological base of reproduction of one copy in which the authenticity of the original is lost, perceiving it another way, not as an end product but as the beginning of new associative images. That relation of serial production we could accept as a view point toward modern and contemporary technologies, not as exclusive systems but as inclusive frames of intense dialogue with particular places.

Just by avoiding the binary opposition of place–non-place and searching for real relations it's possible to recognize contemporary everyday places. Modern architecture has differentiated places on the basis of certain criteria and split their appearances according to before and after; super-modernism swept across spaces and excluded the gaps of authentic places. Based on today's examples of marginal everyday places we can recognize the potential of a direct ratio to the site as open territory and inclusive technology. Visions in which the figures of the past can remain, but from which we are liberated, may free the authenticity of today.

## References

Auge, Marc.1995. *Non-Places, Introduction to an Anthropology of Supermodernity.* London-New York: Verso.

Bakalchev, Minas. 2004. *Housing as an Urban Fragment on the Case of Skopje* PhD diss., University Ss. Cyril and Methodius, Faculty of Architecture, Skopje.

Benjamin, Walter. 1986. *Umetnicko djelo u razdobjlu tehnicke reprodukcije.* Translated by Snješka Knežević. Zagreb: Esteticki ogledi, Skolska knjiga.

Dali, Salvador. 1936. *The Conquest of the Irrational.* http://feastofhateandfear.com/archives/salvador_2.html.

Le Corbusier. 1971. *The City of Tomorrow.* London: The Architectural Press.

Maki, Fumihiko. 1964. *Investigations in Collective Form.* St. Luis: Washington Univertsity, School of Architecture. http://library.wustl.edu/units/spec/archives/photos/maki/maki-part1.pdf.

Rosenberg, Timothy. 2010. One Surrealist a day Blog. *Sunday Dalí: Paranoiac Visage, 1931.* Last modified March 28. http://onesurrealistaday.com/post/480421998/paranoiac-visage.

# CHAPTER FIFTEEN

# ENVIRONMENT CERTIFICATION OF REHABILITATION DESIGN PROJECTS: PUT AND SHU BUILDINGS AS CASE STUDIES

FLORIAN NEPRAVISHTA[1], GERTA VELIU[2] AND RAMADAN ALUSHAJ[3]

**Abstract** | 40% of total energy usage is by the building sector. All the issues discussed toward improving the energy performance of buildings are connected directly with their efficient design. Recently, for environmental certification of design projects, modern computer software techniques have been used. The building energy certificate connects both the implementation of preservation strategies and the utilization of environmentally-friendly solutions to the building sector.

This paper consists of an analysis of environmental factors to minimize the possible negative environmental impact of two existing education buildings. The purpose of this study is to evaluate the environmental parameters achieved through the building rehabilitation project by comparing the conditions obtained with the existing conditions and recommended standards.

Environmental assessment and the use of certification tools are analysed for the selected rehabilitation design projects, those of the Rectory buildings of Polytechnic University of Tirana and Shkoder University. The results of the study will be used to improve design methodology to achieve better energy performance by improving comfort parameters and the efficiency of public buildings.

---

[1] Polytechnic University of Tirana, Faculty of Architecture and Urbanism, Albania florian.nepravishta@fau.edu.al

[2] Polytechnic University of Tirana, Faculty of Civil Engineering, Albania gertaveliu@gmail.com

[3] Polytechnic University of Tirana, Faculty of Mechanics, Albania ralushaj@pklima.com

**Keywords** | Environmental certification, rehabilitation project, energy performance

## Introduction

Global warming is an issue that requires more attention nowadays since we have started to feel its consequences. The building stock in Albania is the country's largest energy consumer. An important issue is to promote energy efficiency, especially in the public sector where performance is particularly poor. Through improving the energy performance of buildings, we might reach the goal of minimizing environmental impact, improving energy efficiency, and reducing economic costs and energy consumption in public sector education facilities.

> *With the best technologies and techniques of environmental certification of design projects, it is possible to evaluate all components of building design. (OECD/IEA, 2010)*

Moreover, we appreciate the certification of the energy consumed in early design projects that have served to implement efficient strategies and environmentally-friendly solutions, to correct the mistakes of the past and their potential future consequences, in order to achieve and secure sustainable development. Environmental assessment is a systematic process that begins by identifying environmental factors, assessing the nature of impacts and raising the necessary profiles for their localization. Environmental analysis includes scanning, monitoring, analysing and forecasting the state of a building.

> *This information was contributed by a model calculation that evaluates a building's energy consumption by local climatic conditions. (Cibse Guide F, 2004)*

## Importance of improving energy efficiency through design

The principles of energy efficiency are supported by the European Commission. The aim is to achieve sustainable energy supply and to meet the target "20-20-20" by the year 2020.

> *Incorporating energy efficiency, renewable energy, and sustainable green design features into public buildings has become a top priority in recent years for facility managers, designers, and other governmental organizations. (European Commission, 2011)*

Activity, hours of occupancy and control settings vary enormously and account needs to be taken of this variability to promote better construction management through improved design. A good management regime can have a major effect on energy consumption. One of the key approaches to low-energy design is to invest in a building's form and enclosure so that the heating, cooling and lighting loads are reduced and, in turn, smaller, less costly heating, ventilating and air conditioning systems are needed (Energy Saving Toolbox, 2002).

## Energy certification and performance of buildings

*Energy Performance Certificates present the energy efficiency of buildings on a scale. An Energy Certificate shows the actual energy usage of a building and the operational rating, and helps one see the energy efficiency based on the annual energy consumption. (Directive 2002/91/EC)*

In preparation of a building energy certificate it is necessary to undertake an assessment of energy consumption. A better-rated building should have less impact in terms of carbon dioxide ($CO_2$) emissions. In existing buildings, energy certification gives the necessary recommendations for the appropriate measures to be taken in rehabilitation design projects (Andaloro, et al., 2010).

The European Directive on the Energy Performance of Buildings will be highly influential in the future. This Directive requires the implementation of:

- *A common methodology for calculating the energy performance of buildings;*
- *Minimum standards for energy performance of new and existing buildings;*
- *Certification schemes when buildings are constructed (Directive 2002/91/EC).*

The method for calculating the energy performance reflects international standards contained in the International Code for Energy Conservation (IECC) and those developed by the American Society of Engineers (ASHRAE, 1992) and the International Organization for Standardization (ISO). The principles of energy efficiency provide a framework for engineers to put the policy into practice. The principles are: reduce demand; understand the building; build in energy efficiency; reduce environmental impact. Therefore, the principles of energy efficiency place great emphasis on management issues (Energy Saving Toolbox, 2002).

In Albania there is an increasing awareness of the need to improve the standards of laws and norms that support energy efficiency in buildings (Nepravishta, 2013), as reflected in Law No. 9379, dated 28.04.2005, "On energy efficiency" and the manual "MTTT Technical analysis works and technological installations".

## Environmental assessment: case study Rectory building of Polytechnic University of Tirana

In the environmental assessment of the first study case, we have analysed the rehabilitation project of the Rectory building of Polytechnic University of Tirana (PUT). The rehabilitation design project represents the organization of facilities based on user requirements, comfort and security while improving existing building heat losses.

To assess the environmental impact of the rehabilitation design project, some calculations of heat losses on each floor were done using the "Termolog EpiX4" software. The efficiency of the heating system was determined by the relationship between energy produced and energy consumed by the system[4].

### Estimation of energy losses

Nowadays innovative approaches for improving thermal performance use integrated assessments of building geometry in combination with the building's exterior and interior properties: walls, windows/doors, floors and terrace systems. The total energy can be calculated using this formula:

$$Q_{tot} = Q_{tr} + Q_{ven}\,(\mathrm{W})$$

Heat loss by transmission from the building envelope (wall, roof and glass) is:

$$Q_{tr} = k\ x\ A\ x\ \Delta t\,(\mathrm{W})$$

where: $Q_{tr}$ – heat loss by transmission (W); k – the heat transfer coefficient ($W/m^2K$); A – heat transfer surface ($m^2$); ΔT – the difference of temperature between two environments ($^0C$).

Heat loss by infiltration is: $Q_{ven} = 0.35 \cdot n \cdot V \cdot \Delta t$

where: $Q_{ven}$ – heat loss by infiltration (W); n – coefficient of volume

[4] This relationship is usually expressed as a percentage and can reach an efficiency of up to 95%.

change of room air per hour (1/h) which in the existing building before rehabilitation was 1.5; V – heated volume ($m^3$); 0.35 – average specific heat of air volume expressed in $Wh/m^3K$.

The ratio of surface to volume for the PUT building is: S/V = 0. 235

The methodology for the consideration of the thermal envelope efficiency takes into account energy efficiency calculated by Gv, the coefficient of volume loss ($W/m^3K$). The calculation for the coefficient of volume losses for the existing building before rehabilitation is:

$$Gv = \frac{Q_{tr}}{V \cdot \Delta t} = 1.86 W / m^3 K$$

where: Gv – the coefficient of volume loss ($W/m^3K$); $Q_o$ – heat losses by transmission from the perimeter of the building; V – geometric volume of the heater location (envelopment of the premise); GD – heating degree days in Tirana, 1132; $t_b$ – interior design temperature (18 $^0$C); $t_j$ – external design temperature (-1$^0$C, Tirana, Albania).

The calculations above show that Gvt for S/V=0.235 in the existing conditions does not meet the thermal requirement for the education buildings. The normative values of the volume coefficient losses with transmission $Gv_{t0}$ and S/V=0.235 according to the respective law must be about 0.52. This means that thermal insulation measures must achieve the required energy efficiency. We have made heat loss calculations through the computer software "Termolog EpiX4". These calculations provide an opportunity to select optimal thermal resistance U-values based on the concept of the thermal performance design of the whole building with identification of optimal thermal conductivity coefficients for all construction components (Fig. 16-1).

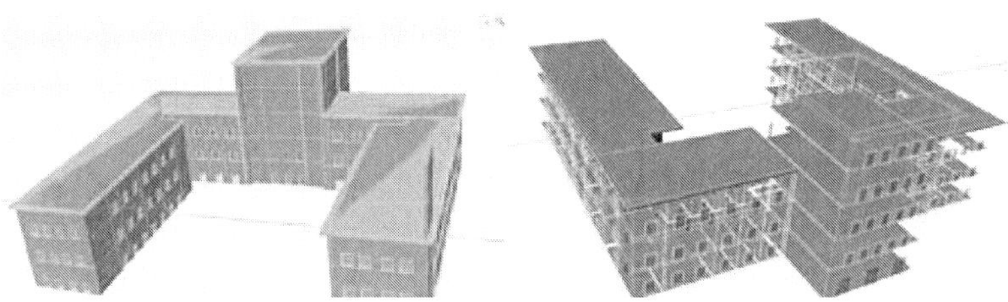

***Figure 16-1***. *Software Interface TERMOLOG (Google TERMOLOG EpiX 4).*

The program can compare two different estimation scenarios. For example, in the case of improving energy efficiency, it compares the existing situation of the building and the rehabilitation design, calculates savings in energy demand and the time and costs incurred for the repair work.

We can compare different design solutions in order to evaluate the proposals more effectively. Through analysis done for 9,954m$^2$, the thermal calculations of the building arrived at the results shown in the chart below (Chart 16-1).

***Chart 16-1***. *Heat losses and comparison of the existing building with the rehabilitation of PUT.*

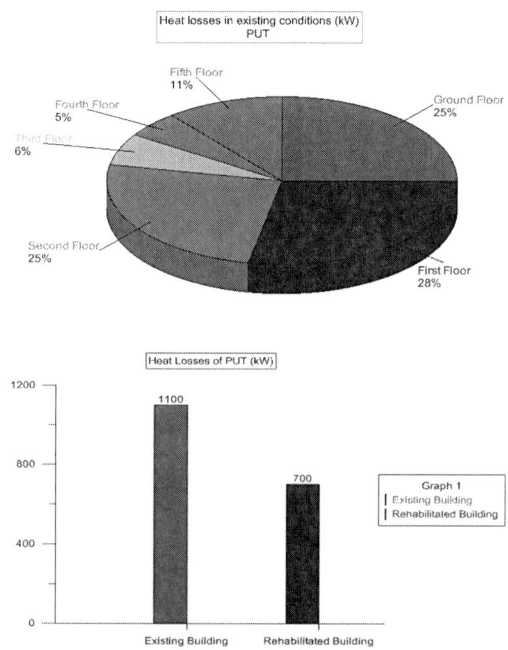

The results taken from the heating load corresponding to the existing situation and after rehabilitation of the building are presented in the following table (Table 16-1):

***Table 16-1***. *Heat losses in existing building of PUT before and after rehabilitation.*

| PUT Rectory Building | Heat Losses (kW) | Area ($m^2$) | Volume ($m^3$) | S/V | $G_{vt}$ ($W/m^3K$) |
|---|---|---|---|---|---|
| Existing building design | 1100 | 9,954 | 42,413 | 0.235 | 1.3 |
| After rehabilitation design | 700 | 9,954 | 42,413 | 0.235 | 0.87 |

## Estimation of energy consumption for heating system

Based on the estimation, the necessary heat capacity for heating after rehabilitation of the Rectory building of PUT is 700 kW with improvement of infiltration such that n = 0.5 (1/h). All the energy needed for heating will be covered by new VRF systems with high energy-efficiency that can reduce the $CO_2$ emissions. VRF systems have not only a high efficiency, with a maximum capacity (COP) of 3.3 to 4.52, but they also have absolutely the highest efficiency in partial load operation that ranges from 5.6 to 6.4 which makes these systems absolute savers of energy consumed (Energy Saving Toolbox, 2002). The indoor air quality is improved with fresh air and ventilation utilizing energy recovery systems with a temperature exchange efficiency of 76%.

The value of energy consumption ($Q_{en}$) for the heating system that is needed to compare the existing building after rehabilitation design is:

$$Q_{en} = Q_{tr} + Q_{ven}\,(\text{kWh})$$

$Q_{tr}$ – heat loss of transmission is:

$$Q_{tr} = GD \times G_{vt} \times 24 \times \frac{V}{1000}\,(kWh)$$

$Q_{ven}$ – heat loss by infiltration (kWh), where the coefficient of volume changes of room air per hour (1/h) n=1.5, is:

$$Q_{ven.} = 1.5 \times 0.34 \times V \times GD \frac{24}{1000} (kWh)$$

*Table 16-2. Energy consumption by heat losses of PUT building.*

| PUT Rectory Building | n (1/h) | $Q_{tr}$ (kWh) | $Q_{ven}$ (kWh) | $Q_{en}$ (kWh/year) |
|---|---|---|---|---|
| Existing building | 1.5 | 1,497,959 | 587,660 | 2,085,620 |
| After rehabilitation | 0.5 | 1,002,480 | 195,866 | 1,198,367 |

## Lighting

The Rectory building of PUT has considerable daylight because of the large surface area of the windows. The artificial lighting is achieved through fluorescent lamps as the most efficient possibility to provide a low value of electricity consumption. The annual consumption of the lighting system is presented in Table 16-3.

*Table 16-3. Annual consumption of lighting system.*

| Type of lighting | No. of lamps | Power (W) | Working Hours | Daily consumption (kWh/day) | Annual consumption (kWh/year) |
|---|---|---|---|---|---|
| CFLs | 607 | 100 | 8 | 485,6 | 140,824 |

## Environmental assessment: case study Rectory building of Shkoder University

The Energy Performance Certificates present the energy efficiency of buildings on a scale.

In the rehabilitation design of the Rectory building of Shkoder University (ShU) in terms of functional and architectural requirements, a range of interventions were proposed such as:

- A common methodology for calculating the energy performance of buildings,
- New thermal insulated walls and terrace,
- Replacement of existing windows and doors with more efficient alternatives,
- Refurbishment of the heating system, fire protection, etc.,
- Improvement of lighting.

## Estimation of heat losses

We calculate the heating load for each floor by applying the above formulas, and the results are presented in Table 16-4.

***Table 16-4***. *Estimation of heat losses.*

| ShU Rectory Building | Heat Losses (kW) | Area ($m^2$) | Volume ($m^3$) | Gvt ($W/m^3K$) |
|---|---|---|---|---|
| Existing building | 245 | 2,410 | 9,640 | 1.27 |
| After rehabilitation | 145 | 2,410 | 9,640 | 0.75 |

We have made heat loss calculations through the computer software Termolog EpiX4. From the existing situation of the building we calculate heat losses of 245 kW and after the rehabilitation we have 145 kW heat losses from the ShU Rectory. All the results found for 2.410 $m^2$ are shown in Chart 16-2, where we can see the heat losses between the two situations of the ShU building.

***Chart 16-2***. *Total heat losses and comparison between two situations of ShU Rectory building.*

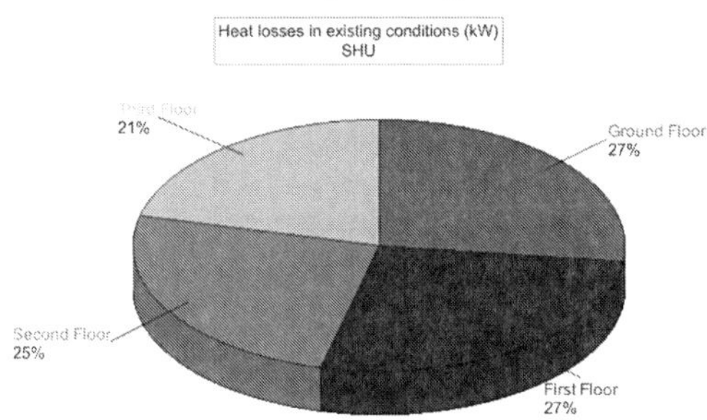


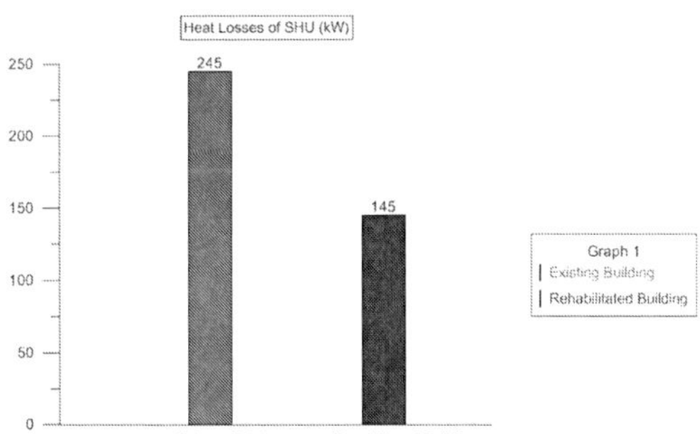

The coefficient of volume losses, Gv, for the existing building before rehabilitation is Gv = 1.27 W/m$^3$K and after rehabilitation design is Gv =

0.75 W/m$^3$K,

where: Gv – the coefficient of volume loss (W/m$^3$K); $Q_o$ – heat losses by transmission from the perimeter of the building; V – geometric volume of the heat location (envelopment of the premise); GD – heating degree days for Shkodra 1392; $t_b$ – interior design temperature (18$^0$C); $t_j$ – external design temperature (-2$^0$C, Shkoder, Albania)

## Estimation of heating energy

Based on the estimation of the heat capacity after the building rehabilitation, the energy needed is 145 kW. The result of energy consumption for the heating system in case of the Rectory of ShU is shown in the Table 16-5:

***Table 16-5***. *Energy consumption by heat losses of ShU building.*

| ShU Rectory Building | n (1/h) | $Q_{tr}$ (kWh/year) | $Q_{ven}$ (kWh/year) | $Q_{en}$ (kWh/year) |
|---|---|---|---|---|
| Existing building | 1.5 | 409,007 | 164,247 | 573,254 |
| After rehabilitation | 0.5 | 241,539 | 54,749 | 296,288 |

This energy will be covered by the VRF systems with a high efficiency, maximum capacity (COP) of 3.7 to 4.52, but they have absolutely the highest efficiency in partial load operation that ranges from 5.6 to 6.4 which makes these systems an absolute saver of energy consumed.

## Lighting

The use of fluorescent lamps for lighting has ensured a higher efficiency of energy use, as well as better comfort in terms of education buildings. Replacing all the incandescent lamps with high-efficiency T5 compact fluorescent bulbs (CFL) has reduced significantly the energy consumption (Table 16-6).

***Table 16-6***. *Annual consumption of lighting system in ShU Rectory building.*

| Type of lighting | No. of lamps | Power (W) | Working Hours | Daily consumption (kWh/day) | Annual consumption (kWh/year) |
|---|---|---|---|---|---|
| CFLs | 167 | 32 | 8 | 42.8 | 15,604 |

## Comparison between the two case studies

The insulation of the external facade of the buildings, floors and terrace is seen as an appropriate measure to take in the building rehabilitation. Comparing the two case studies we note that we have reduced the thermal losses significantly. We have reduced energy consumption by 134 kW for the PUT Rectory building and 72 kW for the ShU Rectory building (Chart 16-3).

***Chart 16-3***. *Comparison of total heat losses and annual energy consumption before and after rehabilitation of two buildings.*

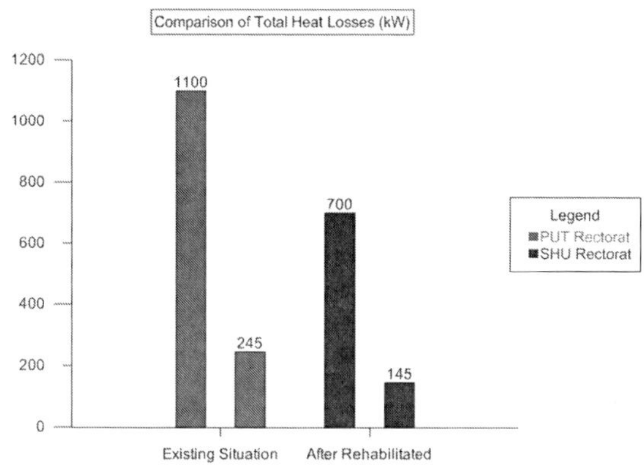


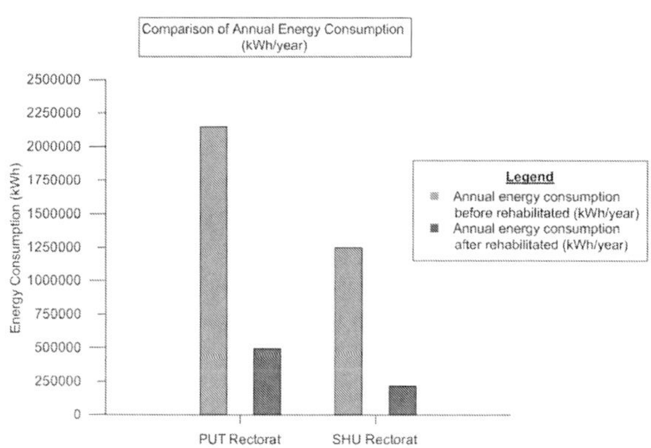


On the chart above, we take an estimate of the significance of the installed systems and make a comparison between these two systems for the buildings studied.

Table 16-7 gives the value of heat losses from the current state of the properties and has defined the energy class of each building.

***Table 16-7**, Energy consumption per $m^2$ year for each building and energy classification.*

| The building | PUT Heat Losses | | ShU Heat Losses | |
|---|---|---|---|---|
| | kW | kWh/m$^2$ year | kW | kWh/m$^2$ year |
| Existing Situation | 1,100 | > 160 | 254 | 160 |
| **Energy Class** | **Class G** | | **Class F** | |
| After Intervention | 700 | 64 | 145 | 52 |
| **Energy Class** | **Class C** | | **Class B** | |

As can be seen from the results in the above table, and based on the energy-efficiency classifications of the buildings according to heat losses in kWh/m2year, the consumption in these buildings descends from class F to class C (Fig. 16-2).

Improving efficiency would lead not only to the improvement of the financial indicators of universities, but would also improve comfort thermo-hygrometric conditions. On the other hand, we also need to emphasize the reduction of GHG emissions' impact on the environment.

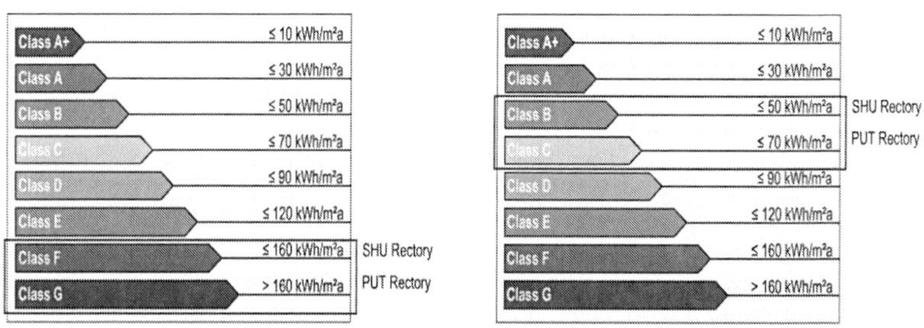


***Figure 16-2**. Classification of energy class in two existing buildings (TERMOLOG EpiX 4).*

## Conclusions

After the environmental assessments of two case studies the Rectory buildings of Polytechnic University of Tirana and Shkoder University, we conclude that the existing situations of both buildings have a low energy classification because of their status and that the main problems that affect heat losses come from air infiltration and thermal insulation. The existing buildings show a discrepancy with the state standards for energy consumption. Through the architectural interventions in the rehabilitating design project (replacement of windows, doors, etc.) the value of heat losses by infiltration is reduced. With the building rehabilitation, we have achieved a lower coefficient of volume losses.

Rehabilitation projects of the existing buildings can reduce and mitigate the negative environmental impact of these structures, as priority issues, by the diffusion and application of design innovations that can lead to more resource and energy efficiency. The environmental assessment of the design projects enables designers to find the best solutions to improve the energy performance of buildings and diminishing the negative environmental impact.

## References

Andaloro, A.P.F., Salomone, R., Ioppolo, G., Andaloro, L., 2010. *Energy certification of buildings: A comparative analysis of progress towards implementation in European countries*, in Energy Policy, 43.

ASHRAE, 1992. *Standard 55: Thermal Environmental Conditions for*

*Human Occupancy*. Atlanta, ASHRAE Inc.

Directive 2002/91/EC of the European Parliament and the Council of 16

December 2002 on the energy performance of buildings. Official Journal

L 001, 04/01/2003 P. 0065 - 0071

Cibse Guide F., 2004. *Energy Efficiency in Buildings*. CIBSE Publication

Department, Page Bros. (Norwich) Ltd., 2004.

Energy Saving Toolbox, 2002: an Energy Audit Manual and Tool.

*Technical Supplement/ Energy Fundamental*, 124-163. Canada.

European Commission, 2011. Energy Efficiency Directive. The 2020 climate and energy package. http://ec.europa.eu/clima/policies/package/

Google TERMOLOG EpiX 4. *Modulo CERTIFICATORE.* Il software modulare di calcolo per progettisti e certificatori http://www.logical.it/software_termolog.aspx.

OECD/IEA, 2010. International Energy Agency. *Policy Pathway: Energy Performance Certification of Buildings*. www.iea.org/papers/pathways/buildings certification.pdf.

Nepravishta, F., 2013. *Rehabilitimi dhe Adaptimi i Ndërtesave.* Monograph, Printing House "Pegi", Tirana 2013.

# CHAPTER SIXTEEN

# GREEN URBAN STRATEGIES IN THESSALONIKI IN THE CONTEXT OF CRISIS

EVANGELIA ATHANASSIOU[1]

**Abstract** | The paper reviews recent developments regarding the urban environment of Thessaloniki. It focuses on the way urban sustainability is constituted in multiple policy levels and pursued in official urban strategies and interventions.

In Greece, in the context of the current crisis, environmental protection is constituted in contradictory and diverting ways in the public realm: on the one hand it is presented as an obstacle to economic growth and on the other it is promoted as the driving force of 'green development'. With regard to urban planning policy, the pursuit of urban sustainability, incorporated into planning law since 1997, is severely challenged by national laws and policies that strive to create a better 'business climate', bypassing environmental controls with a view to making private and public land more attractive to private investment. Nevertheless, the 'green dimension' is prevalent in recent local strategies, urban design projects and interventions in Athens and Thessaloniki. Urban regeneration with an environmental agenda is seen as a means to improve the image of the city thus, making it more attractive to tourists and investors. Notwithstanding its current prevalence in official rhetoric, the goal of urban sustainability is stripped from its social dimension and becomes part of the pervasive urban competitiveness agenda.

The paper presents critically the means and rhetoric of official local strategies aiming to improve the urban environment in Thessaloniki. Small scale interventions aiming at increasing green open space and reducing car traffic form the core of such strategies.

[1] Assistant professor, Department of Architecture, School of Engineering, Aristotle University of Thessaloniki, Greece, e-mail evieath@arch.auth.gr

Subsequently, the paper juxtaposes the claims, rhetoric and practices of recent expressions of the civil society, informal citizens' groups and networks, which increasingly also place different aspects of environmental protection centrally in their agenda. It is argued, that such bottom – up expressions often present composite agendas, which become active fields of synthesis of social and environmental struggles.

**Keywords** | sustainable cities, green urbanism, civil society, urban political ecology, Thessaloniki

## Introduction

The current economic crisis forms a new framework within which the urban environment is lived, regulated, developed, transformed, and discussed. In strictly environmental terms, the interaction of economic, social and environmental processes has resulted in, measurable, environmental alerts, such as the rise of levels of airborne particles due to extensive wood burning in urban homes, and unexpected environmental improvements such as that in traffic congestion due to the decrease of cars circulating in big cities. This paper will not discuss such measurable environmental shifts, inflicted by the crisis, but rather the way environmental protection is approached and the notion of urban sustainability is constituted in the public realm, in the context of crisis.

Within a multi-scalar process of "roll-back" neoliberalisation (Peck and Tickell 2002) that is both triggered and legitimated by the crisis, the whole framework of environmental protection and urban planning, is often projected as a set of concrete obstacles to the development cause and therefore is often relativized or altogether abandoned. The official rhetoric appears to endorse a "post-material" approach to environmental protection and sustainability, at large. In other words, the central government sidelines environmental protection implying that the country is, at the moment, "too poor to be green" (Martinez Alier 1995). On the contrary, it promotes policies that allow bypassing environmental controls and planning restrictions with a view to attracting development on a more flexible environment.

At the same time, a diverting rhetoric is also construed, inspired by the tenets of ecological modernization (Hajer 1995, Spaargaren, 1997 etc.) and "green economy". Ecological modernization, as it consolidated in "sustainable development" and "green economy", suggests that there is no conflict between environmental protection and growth. Quite the contrary, the two can be mutually supportive. The state is expected to facilitate the private sector in tapping the opportunity created by, for example, sustainable management of waste, renewable energy production, or protection of fragile ecosystems, to trigger "green development". Such "greenness" is, at best, disconnected from political and social processes and totally compatible with the promoted array of neoliberal reforms. At worst, it remains a mere smokescreen, offering legitimation to disputed 'fast-track' development.

This contradictory constitution of sustainability is shaped in many policy levels, originating at the level of global economic developments and actually materialising at the level of local urban strategies. The paper traces this multiscalar process as it reshapes the content of urban sustainability and materialises in specific strategies for the urban environment of Thessaloniki.

## A multiscalar process

Sustainable urban development has been incorporated into the framework law, which regulates urban planning in the country since 1997. The 'compact city' model, promoted by the EU since 1990, has been adopted wholeheartedly by planning laws and plans for Greek cities even though they are among the most densely developed cities in Europe. The objective of urban compaction in Greece, however, has not been to increase densities or encourage higher mixture of land-uses within cities, but to abate one of the most notable features of recent urban development, i.e. the diffuse urbanization of peri-urban and non-urban landscapes[2]. All tiers of planning, from the national framework law to General Urban Development Plans of specific municipalities, state their objective to control diffuse urban development in peri-urban areas, areas of ecological interest and agriculture, and limit scattered industrial and tourist development in specific designated areas[3]. Nevertheless, a different direction is actively pursued, in practice, through "growth enhancing reforms" of legislation.

---

[2] See Leontidou L Afouxenidis A Kourliouros E Marmaras E 2007 "Infrastructure – led Urban Sprawl: Mega – Events and Hybrid Peri-urban Landscapes in Southern Europe" in *Urban Sprawl in Europe: Landscapes, Land-use and Policy* edited by C Couch L Leontidou and G Petschel – Held, 71-101, Oxford: Blackwell Publishing.

[3] For instance, the 'compact city' model is adopted by the recently updated Master Plans of Athens and Thessaloniki, which are currently in the process ratification after about 4 and 3 years of public debate respectively.

In the context of the current economic crisis, land, whether urban or non-urban, is seen by the central government as an attractive terrain for private development on the one hand and as an untapped source of immediate revenue on the other. Most eloquently, Law 3894/2010 entitled "Accelaration and Transparency in Strategic Investment", also known as "fast track" Law, forms the central piece of legislation aiming to create a better "business climate" in order to attract capital, to facilitate investment and to re-ignite growth. With the view to reducing bureaucratic delays, it allows, among other things, modification of planning regulations and restrictions stemming from approved Local Development Plans and Urban Layouts. In non-urbanised land, the "fast track law" permits the development of protected areas, such as coastlines, by private companies.

The reform of urban planning legislation is part of the country's obligations as prescribed in the *Second Economic Adjustment Programme for Greece* (2012) and imposed by the European Union, the European Central Bank and the International Monetary Fund. According to the *Programme*, among other "growth enhancing reforms", the government should "review and amend general planning and land-use legislation ensuring more flexibility in land development for private investment, and the simplification and acceleration of land-use plans". To this direction, planning institutions, regulatory frameworks and urban development plans, as developed since the 1980s in the country, are presented at the moment as an impediment to the development of private property, and hence, are by-passed, relativized or altogether abolished. Within the framework of the promoted strategy for national recovery, the existing urban planning framework is first presented as a complicated mechanism causing unnecessary delays and thus hindering growth and then is systematically dismantled. Hence, it can be argued that the very objectives of urban planning are totally transformed: from promoting social equity and environmental improvement through regulating *laissez faire* development, to actively stripping urban and non-urban land from restrictions in order to facilitate private development. Recent changes to landuse regulation as well as abolition of the supervising *Organisations for the Master plan and Protection of the Environment of Athens and Thessaloniki*[4] contribute to this end, accelerating planning processes by reducing control and eliminating social participation.

---

[4] Both Organisations were supervising and coordinating the implementation of the two cities' master and local development plans. They were abolished, along with another 19 public authorities in January 2014, as part of the country's obligations included in the *Second Economic Adjustment Programme for Greece* (2012).

The above transformations are not, of course, a Greek specificity. The pervasiveness of the neoliberal dogma, as it consolidates in different local contexts, dictates policies that destroy existing institutions that safeguard the role of the state –"roll back" neoliberalism- and create new policies and institutions that introduce the market as the sole drive of development, i.e. "roll out" neoliberal policies. Urban policies, along these lines, form a crucial part of the "scalar constitution of neoliberalism" and have been tested in a host of other countries of the capitalist core and periphery.

At the same time, the government is currently seeking to extract quick revenue from unauthorised limitless urbanisation that has persistently undermined the possibility of rational spatial planning, and has had various detrimental environmental effects in landscapes, forests, creeks and beaches. Through successive laws, introduced in the context of the crisis, unauthorised buildings are given the opportunity to become, temporarily, legalized through payment of a fine. Previous legalisations[5] have triggered the phenomenon and produced new generations of unauthorised buildings in periods when urbanisation rates were low and housing needs were not pressing.

Central government policies regarding urban planning and development undermine the pursuit of sustainability, as it has been construed, albeit fragmented and partial, in the national framework of spatial planning in the country. Indirectly, through fiscal policy and investment-promoting legislation, and directly through, deregulating land development and environmental protection, urban diffusion is promoted on flexible landscapes inviting investment and producing revenue. Actively contradicting the goal of urban compaction and the protection of the surrounding landscape from low density unregulated development, diffuse urbanisation is currently seen as a means of collecting revenue and promoting growth.

In a different vein, in recent years, environmental protection and the concept of sustainable development are often activated in the official rhetoric to legitimise policies and projects, attributing them a convenient "green" character. In view of climate change, the interest of the global market focuses on renewable energy production which requires big stretches of land for infrastructure. Towards this direction, large scale investment for solar and wind parks are promoted in Greece, through "fast

---

[5] The last operation of legalisation of unauthorised settlements was launched in 1983. Although there is no official record, there is an estimated increase of unauthorised buildings from 300.000 in 1983 to over 1.000.000, at present (Newspaper *Ta nea*, 19/4/2011).

track" procedures, while their environmental and social consequences are debatable. Degrading landscapes, depleting water resources, appropriating big stretches of land, consuming precious agricultural land and forests, "green" projects often encounter resistance from local communities. Technically their "greenness" is based on the grounds of abating global climate change, while their local impact is often deleterious to local natural processes, catastrophic to landscapes and, to say the least, unsettling to communities. Local development appears to be dictated by the indisputable goal to partake in the global economy, on the one hand, and the state of emergency created by the global, scientifically proven environmental crisis, on the other. The "green", let alone the sustainable, character of such investment remains a veneer.

Summing up, two conflicting approaches to sustainability have been clearly construed in the recent official rhetoric in Greece: one that views sustainability and environmental protection as a "post-material" cause and a hindrance to economic growth and one that ensues the tenets of "ecological modernization", which reconciled economic growth with environmental protection and formed the base for the idea of 'green development'.

## Greening the competitive city

At the level of urban policies, pursued by local authorities, the rhetoric of collaboration between the objective of sustainable urban development and that of urban competitiveness is endorsed and actively promoted. Local urban strategies, plans and interventions for both Athens and Thessaloniki invest on 'greenness", with a view to improving not only the quality of the environment, but also the image of the city to the world.

The logic of collaboration between sustainability and competitiveness is eloquently expressed and pursued in recent strategies and interventions of the Municipality of Thessaloniki. Set within the transient framework of urban planning in the country, municipal initiatives, policies and specific interventions outline a "green" agenda for the city.

First, in an effort to internationalise the city's image and promote a renewed green image to the world, the Municipality bids for European awards and titles and participates in European networks with an environmental agenda. In 2011, the city became a "golden" member of the *European Mobility Week*, which is a network of cities, under the auspices of the European Commission, promoting sustainable mobility through temporary events and permanent arrangements. The same year, the Mayor signed the *Covenant of Mayors* committed to sustainable energy and three, years later, in April 2014, an *Action Plan* was submitted, presenting concrete action towards reducing the city's $CO_2$ emissions by 20%, until 2020. Finally, the city bade for the *European Green Capital 2014* award, which it did not manage to win.

Second, notwithstanding recession and shrinkage of state funding, there has been increased interest on urban public space, in the form of architectural competitions and small-scale interventions. More specifically, five architectural competitions were launched between 2011 and 2013, regarding redesigning public squares or designing totally new ones. They all had environmental requirements in their briefs. Three competitions eloquently stated their objective of "projecting a renewed image of the city" and one clearly included improving quality of life and revitalizing local economy, through increasing competitiveness, among its objectives.

In addition, there has been a considerable amount of small scale low budget interventions, planned, designed, financed and built by the Municipality. Those had a clearly expressed intention to reclaim space back from car traffic to the benefit of pedestrians and cyclists, discouraging car use and reconciling car use with pedestrian movement, through traffic calming arrangements. Such small scale urban interventions have originally been located in the city centre. At a later stage, they were planned and materialised in off centre neighbourhoods.

Finally, the Mayor stresses the importance of the civil society in urban management and has initiated a new relationship with local expressions of the civil society. A number of events have been organised or supported by the Municipality seeking to encourage citizens' involvement and participation for example in maintaining green spaces, through organised tree planting, playground or bench adoption. The Municipal council and the Mayor personally took a stance against the centrally planned privatization of the water utility, endorsed and supported the referendum that was organised by citizens' movements[6]. They also reacted against the recently announced privatization and development of two abandoned military camps within the city, which are locally seen as the last opportunity for the city to increase its tiny percentage of green space per capita.

---

[6] A referendum posing the question of privatization of the water utility company was organized by two citizens' movements on 18 May 2014, in parallel to local elections. The referendum was carried out exclusively by volunteers. Municipality of Thessaloniki along with other municipalities of the metropolitan area supported the referendum. See http://sostetonero.blogspot.gr/ and http://www.136.gr .

It is clear that 'greenness' is part of the dominant agenda in the city. Sustainable mobility and increased amount of green spaces are the core issues, while less projected issues of urban planning and urban metabolism are sidelined. The goal of an improved urban environment is harmoniously combined with the goal to internationalise the city, to promote a renewed image to the world and to make it more attractive to tourists and investors.

However, this "green" turn of the Municipality, pursues a relativized and reduced version of the multi-facetted concept of sustainability, which prioritises strictly environmental conditions and the urban economy, while marginalizing social and political aspects of the urban environment. The new green identity of Thessaloniki, needs to be considered in conjunction with recent initiatives of the Municipality, which surrender maintenance of specific central parks and squares to corporate control, intensify clearance operations of "undesirable" users of public space, such as migrant street vendors, and ban political campaigns from central public spaces to protect local businesses from the disruption (Athanassiou, 2012). Sustainability is stripped from its social dimension and thus reduced to partial, visual and non – political "greenness".

Without seeking to deny the potentially positive impact of certain recent strategies in, measureable, strictly environmental aspects of the city, this paper highlights the following:

- Green strategies in Thessaloniki are set within the framework formed by multiscalar processes, emanating from the global pervasiveness of the neoliberal dogma, its European manifestation in the context of the crisis, and its national materialization in aggressive reforms regarding the urban environment which undermine basic dimensions of urban sustainability and environmental protection.
- Green initiatives in Thessaloniki are seen as contributing to an attractive image of the city, thus, promoting urban competitiveness that is projected as an indisputable non-political objective for the city, in the context of global economy.
- Urban sustainability is reduced to strictly measurable environmental aspects of "greenness", which are construed as non-political and indisputable requirements of technical nature, dictated by the state of emergency created by the climate change and global environmental crisis.

## Fields of synthesis

The separation of environmental protection from human social and political processes is not specific to Greek environmental politics, at national and local level. It has its roots to the very constitution of the environmental discourse and the idea of urban sustainability. In terms of how the city is perceived in relation to nature, the sustainable cities discourse has not departed from what Smith (2006, viii) calls 'the intellectual shibboleth of high capitalist period' i.e. the separation of nature and society and the subsequent understanding of the city as the manmade opposite to nature. The city is seen as a consolidation of static manmade properties, an un-natural 'thing', on which green policies, dictated by a separate entity of superior value called "nature", need to be applied. The state of emergency created by the climate change and the global environmental crisis and the imperative to partake in the global economy form the indisputable framework, in which such policies are vindicated.

A more composite and fruitful approach is developed by some of the expressions of the civil society that have been triggered by the crisis. Urban movements, solidarity groups and networks often place aspects of the urban environment centrally in the agendas. Their approaches, scope and scale of reference in the city differ. Some constitute organized resistances against local issues, such as the development of specific publicly owned plots of urban land, or citywide issues, such the privatization of water utility company. Others seek to promote or protect a particular aspect of life in the city, most commonly, green spaces, and pedestrians' and cyclists' rights, that is undermined or put under threat in the contemporary urban condition.

Many initiatives in Thessaloniki and big cities across the country constitute more than resistances to specific urban adversities, policies or development projects. They pursue composite and genuinely open agendas, which transcend traditional boundaries between social struggles and the quest for environmental protection. Their members demand more than attractive streets, bicycle lanes or green parks. Public space and urban green are understood as active fields of synthesis of environmental, cultural, social and political processes and not as attractive and environmentally sustainable voids. Collective farming of vegetables in abandoned military camps in Thessaloniki is an example of such initiative. Beyond producing fresh locally grown food, the collective seeks to protect public land from private development. At the same time they actually practice a new lived relation between citizens and urban green spaces.

Most important, such urban farmers reinvent ways of creating urban communities in the contemporary metropolis, in conditions of crisis.

In such fields of synthesis, the binary between city and nature is questioned, and "green" becomes a lived process, rather than a technical fix. They constitute a fruitful ground on which a different environmental politics (Loftus, 2013) and "an ecologically enriched public realm" (Gandy, 2006) is actually explored.

## References

Athanassiou, E. 2012 " 'What's up park' or the political ecology of green spaces in Thessaloniki, *3rd Conference of Spatial Planning and Regional Development*, Volos 27-30 September, 117 (book of abstracts ISBN 978-960-6865-51-0 and cd ISBN 978-960-6865-52-7) (in greek).

Covenant of Mayors committed to local sustainable energy. Accessed 20 February 2014 http://www.covenantofmayors.eu/index_en.html.

European Commission (2012) *The Second Economic Adjustment Programme for Greece*. Accessed 20 February 2014
http://ec.europa.eu/economy_finance/publications/occasional_paper/2012/pdf/ocp94_en.pdf.

Gandy M., 2006. Urban nature and the ecological imaginary. *In the nature of cities: Urban Political Ecology and the Politics of Urban Metabolism*, edited by N. Heynen, M. Kaika, E. Swyngedouw, 63-73, London: Routledge.

Hajer M A 1995 *The politics of environmental discourse: Ecological Modernization and the policy process*, Oxford: Clarendon Press.

Law 3894/2-12-2010 Acceleration and transparency in implementation of strategic investment (Government Newletter ΦΕΚ 204 2/12/2010 A).

Leontidou L Afouxenidis A Kourliouros E Marmaras E 2007 "Infrastructure – led Urban Sprawl: Mega – Events and Hybrid Peri-urban Landscapes in Southern Europe" in *Urban Sprawl in Europe: Landscapes, Land-use and Policy* edited by C. Couch L. Leontidou and G. Petschel–Held , 71-101,Oxford: Blackwell Publishing.

Loftus A, 2013 *Everyday environmentalism: creating an urban political ecology*, London: University of Minnesota Press.

Martinez Alier J 1995 "The environment as a luxury good or 'too poor to be green'" *Ecological Economics*, 13 pp1-10.
http://is.muni.cz/el/1423/jaro2011/HEN581/um/Martinez-Alier_Ecological_economics.pdf

Peck J. Tickell A 2002 'Neoliberalising Space' in *Spaces of Neoliberalism: Urban restructuring in North America and Western Europe* edited by N. Brenner and N. Theodore, 33-57, Oxford: Blackwell.

Peri-Urban farming (PERKA) Accessed 20 February 2014 http://perka.org/

Smith N 2006 "Foreward" *In In the Nature of Cities: Urban Political Ecology and the Politics of Urban Metabolism* edited by N. Heynen M. Kaika E. Swyngedouw, 1-20, New York: Routledge.

Spaargaren G 1997 *Ecological Modernization of Production and Consumption*, Wageningen: Wageningen University.

# CHAPTER SEVENTEEN

# INFLUENCE OF GLASS COMPONENT JOINTS ON THE STRUCTURAL GLASS FAÇADE DESIGN

ALEKSANDRA KRSTIC-FURUNDZIC[1], TATJANA KOSIC[2] AND JEFTO TERZOVIC[3]

**Abstract** | There is a strong relationship between the type of glass component connections and the structural glass facade design. The objective of this paper is to study the different solutions to the design of the connections between elements of structural glass facades. Influenced by the load, in relation to the structural system, a glass layer may be exposed to different types and intensities of stress, which inevitably affect the design process in terms of selecting the type and dimensions of glass elements. In order to evaluate the impact of the design of component joints on the facade structural system, and hence its design, some hypothetical models of structural glass components joints are created and numerical analyses are carried out. The methodological approach entails three steps: creation of a typology of joints, which includes form and assembly types; design of hypothetical models of structural glass facade components joints; and examination of the structural stability of hypothetical models.

**Keywords** | Structural glass, form types, assembly types, structural system concept and stability.

---

[1] PhD, Full Professor, Faculty of Architecture, University of Belgrade, Belgrade, Serbia, akrstic@arh.bg.ac.rs

[2] PhD Candidate, Teaching Assistant, Faculty of Architecture, University of Belgrade, Belgrade, Serbia, tkosic@arh.bg.ac.rs

[3] PhD Candidate, Teaching Assistant, Faculty of Architecture, University of Belgrade, Belgrade, Serbia, terzovic@arh.bg.ac.rs

## Introduction

The objective of the paper is to study the influence of glass component joints on structural glass facade design. Facade design includes the design of different types of connections, the characteristics of which influence the concept of the structural system. The paper is intended to serve as a state-of-the-practice summary regarding the connection types of structural glass facades, and in that sense, the paper is structured. The typology of structural glass component connections is based on the fact that the interdependence of structural glass component types, connection place and technology is crucial. Based on connection types, the structural system concept and stability of structural glass facade panels are analysed. The diversity of structural glass panel connections gives the potential for different options with regard to facade transparency and a building's appearance.

The methodological approach entails three steps:

- creation of connection design typology,
- design of hypothetical models of structural glass panel connections, and
- examination of the structural stability of hypothetical models.

The design of hypothetical models is carried out according to defined connection types. The stability of glass panels in relation to various connection types (linear-adhesive or with metal or glass sections, and dotted-bolted or adhesive) is examined through stress and deformation analyses.

> *"The brittleness of glass presents a major challenge when connecting glass construction components. Connection techniques should be based on a glass-conscious construction kit of various solutions designed to suit different stress conditions, which can be combined and modified. In all cases, there must be a uniform loading transfer between glass and connecting elements by means of suitable intermediate layers... The load transfer layer should combine a low modulus of elasticity similar to that of glass with good durability and as high a compressive strength as possible... The form of loading transfer has a great effect on the stress distribution in brittle glass" (Wurm 2007, 85-86).*

The paper indicates various approaches to the design of connections for structural glass facade panels and might be helpful to architects and engineers.

## Connection design typology

Considering the connection types in the case of structural glass facades, the following options can be recognized:

- glass component connections, and
- glass and other structural component connections.

*Table 1. Connection types of structural glass facade components.*

| | | Connection place and technology | | | | |
|---|---|---|---|---|---|---|
| | | On the corner | Along the edges | | Within the surface | |
| | | Point support (bolted) | Point support (bolted, locally clamped, friction-grip) | Linearly supported (adhesive/metal sections) | Point support (bolted, friction-grip) | Linearly supported (adhesive/metal sections) |
| Types of glass structural component | Linear | | 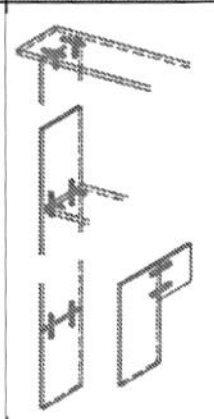 | 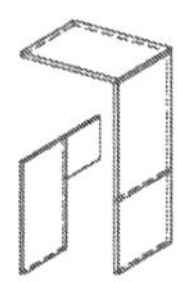 | 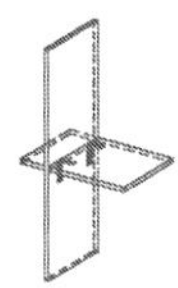 | 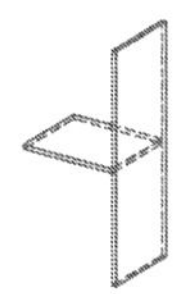 |
| | Linear-plane | 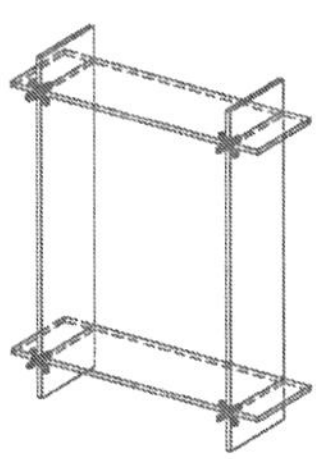 | 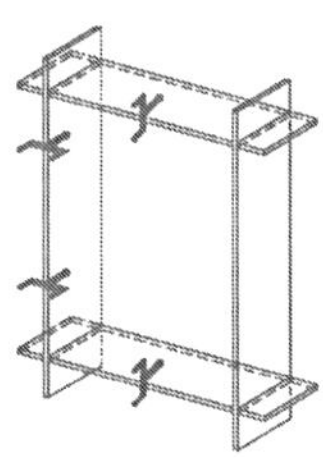 | | 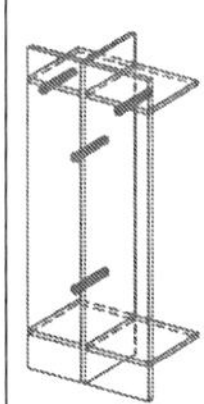 | 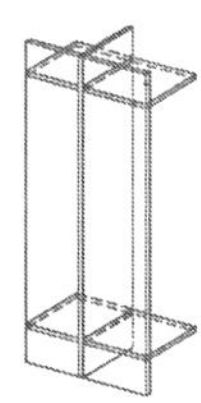 |
| | Plane | 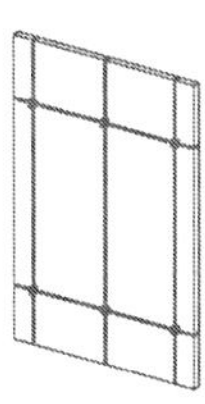 | 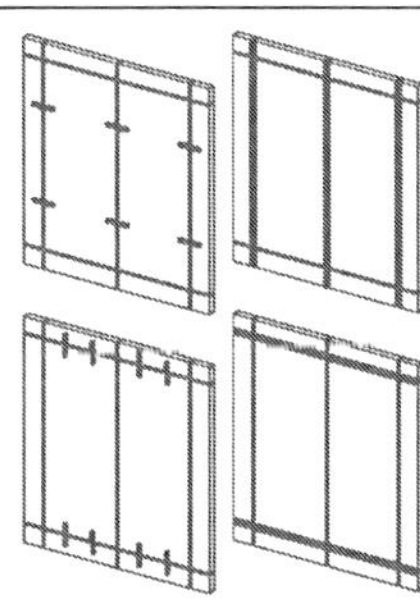 | | | |

The mentioned options set out specific requirements in terms of component types and materials, connection place and technology.

> *"In case of mutual connections of glass facade components, classification can be specified according to:*
> - *glass facade component types (connections between linear components, linear and plane components and plane components),*
> - *connection place (on the corner, along the edges and within the surface), and*
> - *connection technology (dotted-bolted or adhesive and linear-adhesive or with metal or glass sections) as shown in Table 1" (Krstic-Furundzic et al. 2012, 895).*

Also, construction of structural glass facades includes the design of glass and other structural component connections which can be classified according to structural component types (connection with the roof, with the foundation and with the gable) and according to component materials (glass and concrete, glass and wood, and glass and metal).

## Design of hypothetical models of structural glass panel connections

Selection of the connection type between glass facade panels and the building structure, respectively supporting the facade substructure, directly affects the static system of glass facade panels, and still affects the form and visual identity of the facade (Table 2). The connection of the glass facade panels with the building's structure is enabled by the facade substructure, which, in a static sense for the glass panels represent linear and point supports. In that sense, the following hypothetical models of structural glass facade panel connections are selected:

1. Linear connection by using horizontal sections: embedded-invisible or visible horizontal sections (Fig. 1);
2. Dotted connection: on the level of the floor structure, or on the level of the floor structure and half the height of panels (Fig. 2);
3. Linear connection by using frames: visible sections or embedded-invisible horizontal and visible vertical sections (Fig. 3);
4. Linear connection by using vertical sections: embedded-invisible or visible vertical sections (Fig. 4).

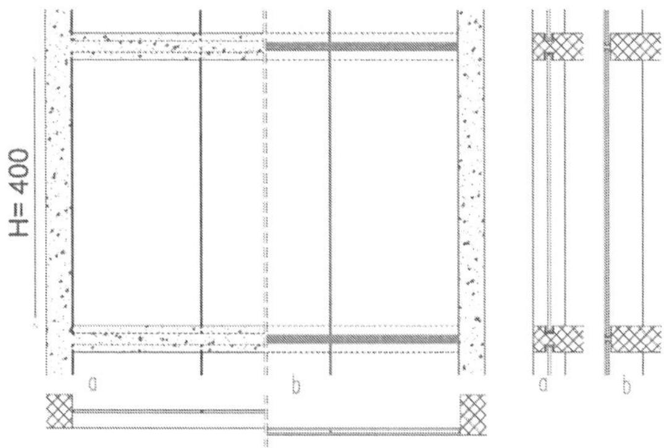


*Fig. 1. Glass panels supported by horizontal sections (a. embedded, invisible; or b. visible horizontal sections) – Model 1.*

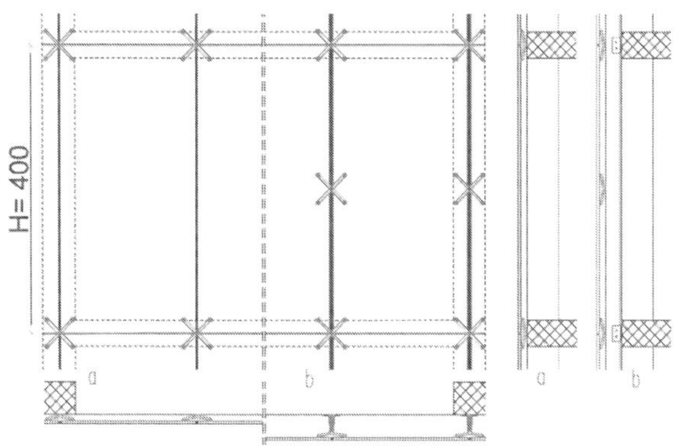


*Fig. 2. Glass panels supported by point joints (a. on the level of the floor structure; or b. on the level of the floor structure and half the height of panels) – Model 2.*

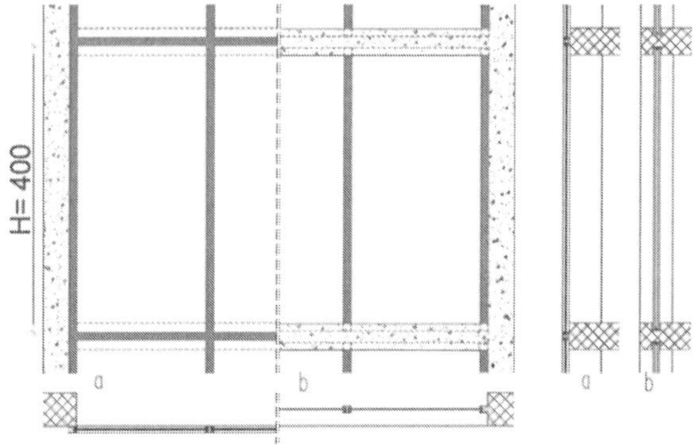


*Fig. 3. Glass panels supported by frame (a. visible sections; or b. visible vertical and embedded, invisible horizontal sections) – Model 3.*

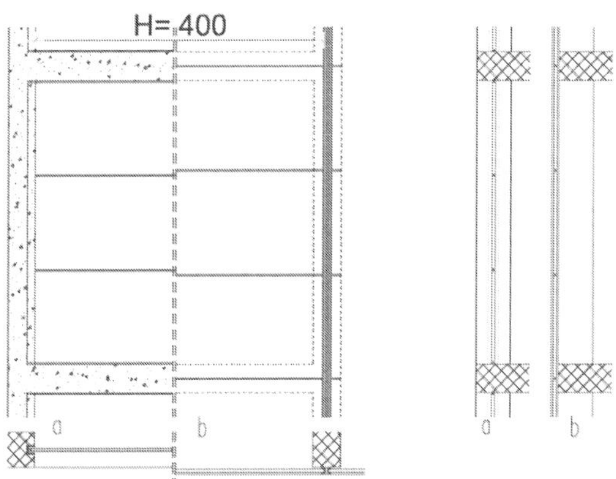


*Fig. 4. Glass panels supported by vertical sections (a. embedded, invisible; or b. visible vertical sections) – Model 4.*

*Table 2. Connection types and visual identity of the facade (©A. Krstic-Furundzic).*

| | **Visible sections** | **Embedded-invisible sections** |
|---|---|---|
| **Glass panels supported by horizontal sections** |  |  |
| **Glass panels supported by frames** |  |  |
| **Glass panels supported by vertical sections** |  |  |

| *Glass panels supported by point joints* |  |  |
|---|---|---|

## Influence of connection type and facade design on static characteristics of structural glass facade panels

A variation of the number of supports and their mutual arrangements provides a variety of connection options for the design of glass facade panels in terms of shaping and material consumption. The computer models of facade glass panels which are 2 m wide and 4 m high (Models 1-3), as well as 4 m wide and 1.33 m high (Model 4), which are loaded by wind force perpendicular to the plane, are created and analysed. For the models, various panel supports were predicted: linear supports only on shorter opposite edges (relying only on the floor structure, or supported only by vertical structural building elements); point supports on panel corners or on panel corners and half the height of panels ("spider" facade); and linear supports throughout the whole scope of the panel, supported by the panel frame (relying on the floor structure or vertical load-bearing elements of the facade substructures). Each of these support types influences the connection requirements. Usually, the supporting option can be clearly identified on the facade in terms of design, and it shows a direct connection between the structural system of the facade and its visual identity.

The analysis was done by software Tower 6.0 (Radimpex Software 2014), based on the Finite elements method which is an assemblage of many mathematical entities (finite elements) as an approximate representation of geometry, material properties, connectivity, and their spatial relationships.

> *"In computer models, glass elements are designed according to their characteristics, which are described by adequate module of elasticity, Poisson's coefficient, volumetric weight, and thermal coefficient" (Krstic-Furundzic et al. 2013, 48).*

## Static analysis of computer models

Glass facade panels rely on the supporting structure of the building, usually along the edge of the floor structure, via elements of the facade substructure. In case of linear (continuous) facade substructure elements connected to the building structure, they withstand negligible stress and all the external load is transferred to the floor structure of the building through shear force in the cross section of fasteners (bolts and anchors), elements by which the facade substructures are attached to the building structure.

In the case that vertical elements of the facade substructure are omitted, the facade glass panel corresponds to the static system of a simple beam. This panel acts as orthotropic plate, carrying in only one direction, since its two edges are free and do not have any contact with the facade substructure or the building structure. The span of the vertical facade glass panel, linearly supported along horizontal edges, consequently, is equal to the storey height of the building. On the contrary, the span of the facade glass panel, linearly supported along vertical edges, is equal to the distance-grid of vertical elements of the facade substructure or of the building structure.

For computer models, the net weight of the glass panel (load case I) and equally divided surface load of intensity of 1.00 kN/m$^2$ (load case II), simulating wind effect, are taken into account.

In the first case, Model 1, the flat panel of 20 mm thick toughened glass, linearly supported along horizontal edges (Fig. 1), is analysed. The results show that the stresses are within the allowed limits, but the deformations are unacceptable because their values are much above permissible limits, as shown in Figure 4.

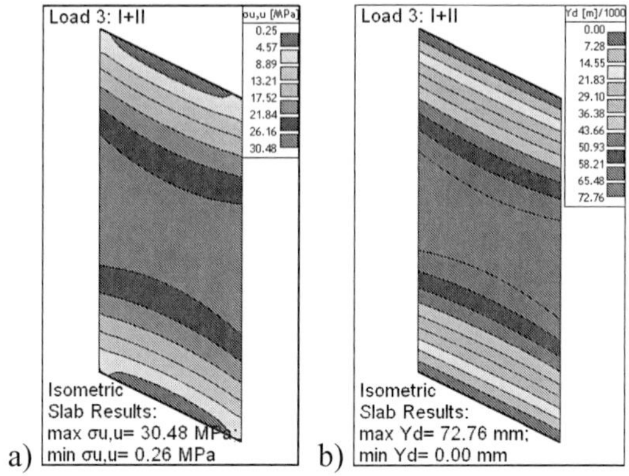


*Fig. 4. Normal flexural stresses (a) and deformations (b) of the 20 mm thick flat glass panel, linearly supported along its horizontal edges by the floor structure (storey height = 4 m).*

Acceptable results of deformations for the same storey height, i.e. the span of the glass panel and the same load conditions, are obtained by using laminated glass panels that are twice as thick (40 mm), as shown in Figure 5.

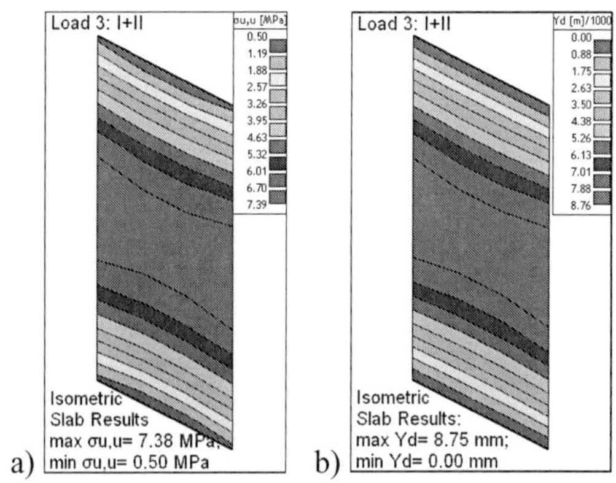


*Fig. 5. Normal flexural stresses (a) and deformations (b) of the 40 mm thick flat glass panel, linearly supported along its horizontal edges by the floor structure (storey height = 4 m).*

This example of the supporting system of the facade glass panel shows that avoiding vertical elements of the facade substructure, respectively avoiding the vertical division of the glass facade surface, dramatically affects the use of material-glass and the facade's appearance.

According to Wurm, pane or plate structural glass elements can be supported on their edges, corners or surface at points or on their edges linearly or by a combination of the two methods, and the form of loading transfer has a great effect on the stress distribution in brittle glass. Also, Wurm considers that point or unevenly distributed load transfer produces concentrated load effects and do not make efficient use of the glass cross section.

> *"The stress at the support points can be up to three times the general stress level, making it usual to have to use thermally treated glass point fixings that are able to accommodate construction tolerances well but the engineering design of the support points is complex" (Wurm 2007, 86).*

For the purpose of this research, models of glass panels supported by point-dotted connection, Models 2a and 2b, are created. Supporting points are placed only on corner areas of the glass panel, on the level of the floor structure (Fig. 2a) or on corner areas and half the height of the glass panel (Fig. 2b). Due to the fact that support points are placed at a certain distance from the panel edge, thereby forming the static system of a beam with minimal overhangs on both sides (Fig. 2a), as well as due to the reduced span of the beam, stresses and deformation values are less than in the previous case – the panel is linearly supported along horizontal edges, as shown in Figure 6.

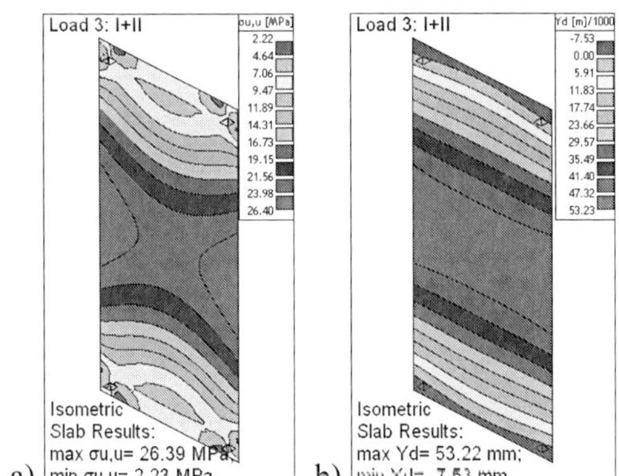


*Fig. 6. Normal flexural stresses (a) and deformations (b) of the 20 mm thick flat glass panel, supported by point connections on its corners on the level of the floor structure (storey height = 4 m).*

By applying point supports on the corner areas of the 20 mm thick glass panel, which are directly connected to the building structure, satisfactory results in terms of stresses and deformations cannot be obtained. In order to achieve the values within the allowed limits, it is necessary to increase the thickness of glass panel (40 mm), as shown in Figure 7.

In the case where new support points are arranged at half the height of the panels, stresses and deformations are reduced. However, this requires the use of the vertical element of the facade substructure in order for the point supports at the middle of the storey height to be placed.

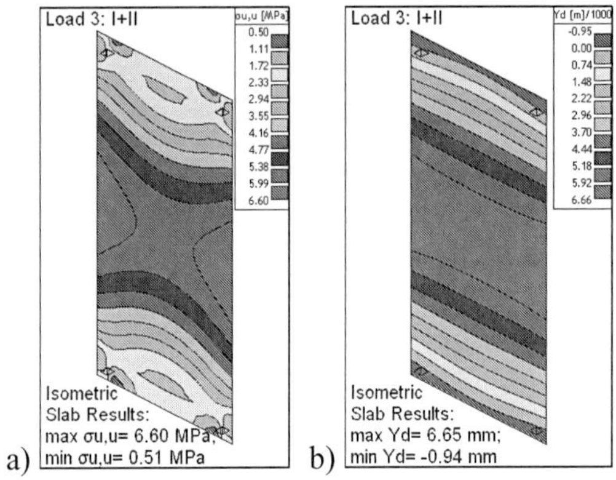


*Fig. 7. Normal flexural stresses (a) and deformations (b) of the 40 mm thick flat glass panel, supported by point connections on its corners on the level of the floor structure (storey height = 4 m).*

In terms of the static system, the vertical elements of the supporting substructure of the facade are the simple beams due to the supports on the level of the floor structure. Usually, the simple beam is represented as a linear support, the span of which is the storey height, which is loaded with a uniformly distributed load. If the simple beam is loaded with a concentrated force in the middle of the span, it retains the same construction concept of glass facade panels as in the previous example of point supports on corner areas (facade panels entirely supported by point supports).

In the following examples, the model of facade glass panels, supported by point connections on their corners on the level of the floor structure and on half the height of the glass panel by vertical sections of the supporting substructure (Fig. 2b), is created and analysed.

The results of the stresses and deflections of facade panels that are 20 mm thick and 4 m in height are shown in Figure 8.

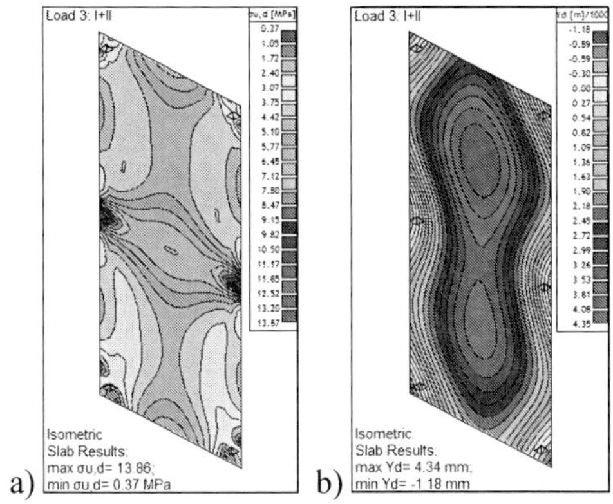


*Fig. 8. Normal flexural stresses (a) and deformations (b) of the 20 mm thick glass facade panel, supported by point connections on its corners on the level of the floor structure and on half the height of the glass panel (storey height = 4 m).*

The results show very good values for the stresses, especially for deflections, enabling rationalization by reducing the panel thickness (Fig. 9).

The vertical element of the facade's supporting substructure receives the horizontal reaction of the point support of the facade glass panel in the form of concentrated force perpendicular to the direction of its longitudinal axis. This implies that the vertical element of the supporting substructure is not only loaded by axial force due to the weight of the facade panel, but also by the bending moment, with respect to which calculations must be taken into account. The intensity of the horizontal reaction of the point support is $R_2 = 2.21$ kN (Fig. 10a). Since one vertical element of the supporting substructure carries two adjacent facade panels, the concentrated force in the middle of the span will be of an intensity equal to twice the value of the reaction force:

$$P = 2 \times R2 = 2 \times 2.21 \text{ kN} = 4.42 \text{ kN}$$

The values of the stresses and strains on the vertical element-section of the supporting facade substructure are within the permissible range, if the cross section of the vertical aluminium alloy section has the external dimensions of 60 x 250 mm, and a wall thickness of 4 mm (Fig. 10b, c).

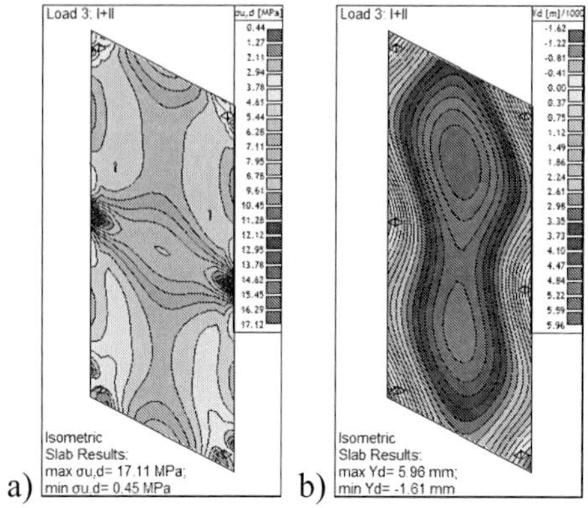


*Fig. 9. Normal flexural stresses (a) and deformations (b) of the 18 mm thick flat glass panel (rationalised), supported by point connections on its corners on the level of the floor structure and on half the height of the glass panel (storey height = 4 m).*

It can be noted that the rationalization of the thickness of the facade glass panel requires the introduction of a vertical element of the facade's bearing substructure, which has the decisive impact on the facade's design.

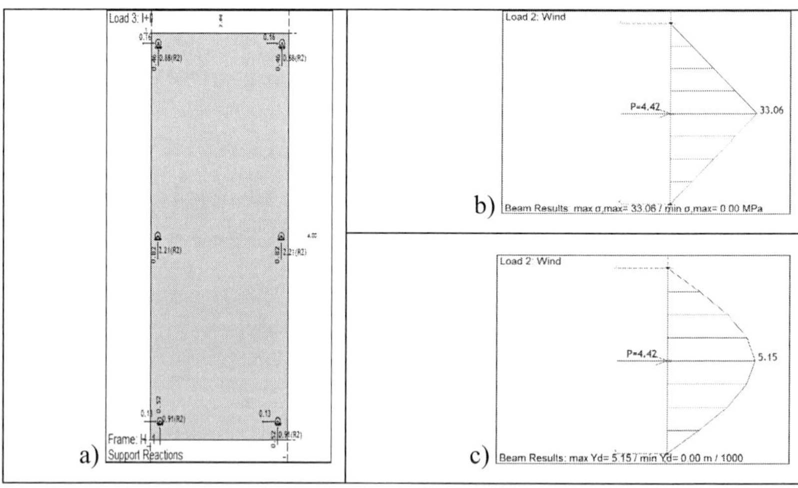


*Fig. 10. Supporting reactions of the glass facade panel (a), stress-bending moments (b) and deflections (c) of the aluminium alloy vertical element-section of the facade-bearing substructure.*

By using the vertical and horizontal elements of the substructure, which are in a static sense the linear supports, it is made possible for facade glass panels to rely on the whole scope of the panel, forming the static system of a plate carried in two orthogonal directions by frames – Model 3 (Fig. 3a, b) – whereby the direction of the shorter span is more burdened. If the distance between the vertical elements is smaller than the storey height, and there is no horizontal division of the glass facade's surface, the horizontal direction is the direction of the dominant strain on the glass panels. In this case, the vertical substructure elements are the main linear glass panel supports forming the static system of a simple beam and they need to be specially treated by static analysis. In the case of the same geometry and thickness of glass panel (20 mm), as in previous cases, Figure 11 shows how stresses and deformations of the facade glass panel are minimized due to the linear supports throughout the whole scope of the panel.

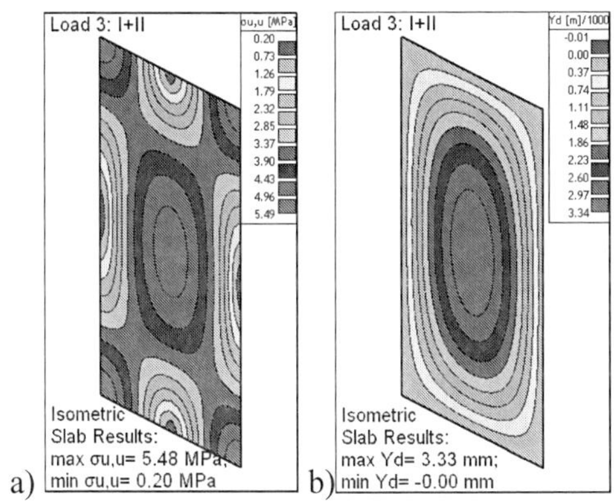


*Fig. 11. Normal flexural stresses (a) and deformations (b) of the 20 mm thick flat glass panel, linearly supported on its edges by vertical and horizontal sections of the facade substructure (storey height = 4 m).*

The horizontal division and orientation of the facade glass panels without the horizontal elements of the load bearing facade substructure, in the case of Model 4, imply that the glass panels only rely on the vertical bearing substructure. In static terms, facade glass panels are simply supported beams, and their span is equal to the mutual distance of vertical elements-sections of the load bearing facade substructure (Fig. 4).

A facade glass panel supported by vertical elements represents the beam in the range of 4 m, the cross section of which is rectangular with a typical height that is more than 30 times greater than its width. Although the dominant load is the wind, perpendicular to the plane of the panel, the glass panel is also strained by the bending moment under the influence of the gravity load of its own weight.

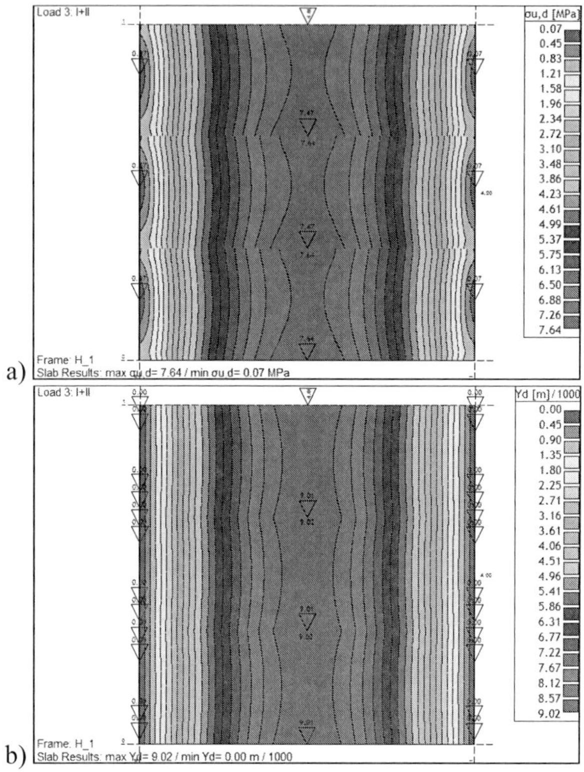


*Fig. 12. Normal flexural stresses (a) and deformations (b) of 40 mm thick flat glass panels, linearly supported by vertical sections (panel width/range = 4 m).*

Figure 12 shows the stresses and deformations of facade glass panels which are 40 mm thick and supported by vertical sections.

Due to the bending moment, on the rectangular cross section of the glass panel, the pressure occurs on the upper edge and tension on the lower edge, while the neutral line is at half the height of the cross section. It is a bracing force of the same intensity and direction, but of the opposite sign (Fig. 13).

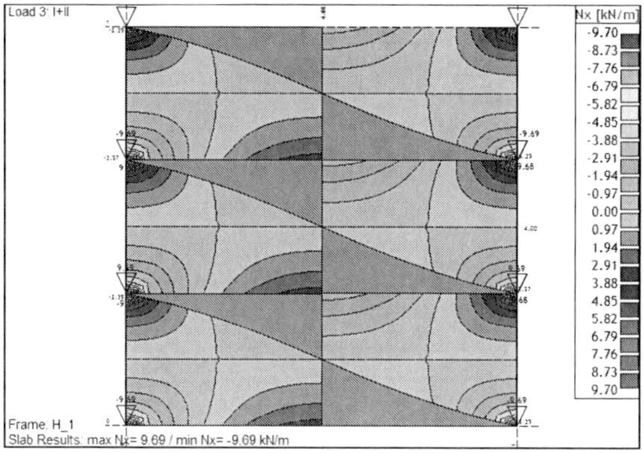


*Fig. 13. Axial forces (Nx) of 40 mm thick flat glass panels, linearly supported by vertical sections (panel width/range = 4 m).*

The design of structural glass facades and the rationalization of the thickness of facade glass panels should not significantly influence the primary structural concept, but it is necessary to consider that vertical elements of the load bearing facade substructure are supported by the floor structure.

Vertical elements-sections are the linear supports of facade glass panels. They are loaded by an evenly distributed wind load which is perpendicular to their longitudinal axis, as well as by the weight of facade panels in the form of axial (Nx) forces.

The intensity of both reactions of the linear support of the facade glass panel amounts to 1.5 kN/m (Fig. 14a). Since one vertical element of the supporting substructure carries two adjacent facade panels, the intensity of the load is 3.0 kN/m, according to which the vertical element of the facade bearing substructure, made of aluminium alloy, has to be designed.

The values of the stresses and strains of the vertical element-section of the supporting facade substructure are within the permissible range, if the cross section of the vertical aluminium alloy section has the external dimensions of 60 x 250 mm, and wall thickness of 4 mm (Fig. 14b, c).

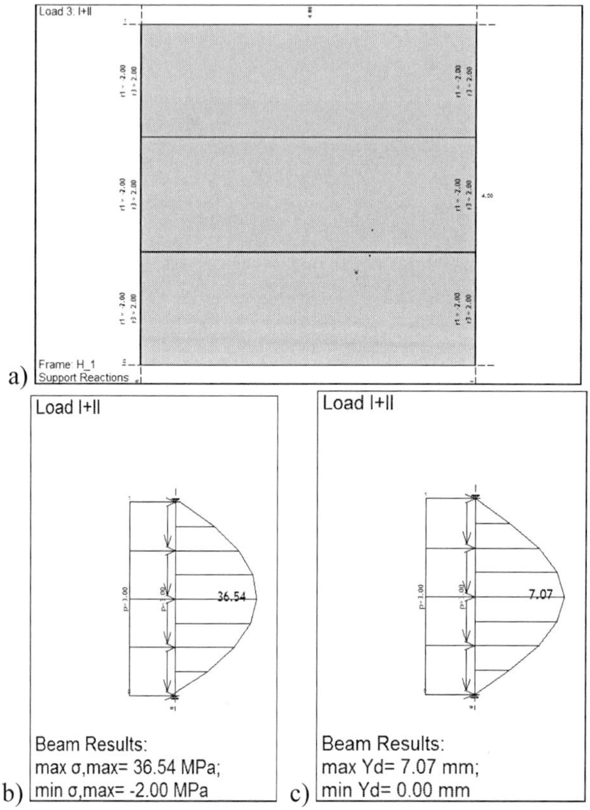


*Fig. 14. Supporting reactions of glass facade panels (a), stress-bending moments (b) and deflections (c) of the aluminium alloy vertical element-section of the facade bearing substructure.*

## Conclusions

From the standpoint of the rationality of the construction, the setting of vertical elements of the facade substructures is crucial, whether they are visible on the facade or embedded, invisible – placed behind the glass layer; whether they are metal or glass elements, whereby this implies that the glass panel is supported on the level of the floor structure. If the vertical substructure element is predicted, there is no significant difference in the stresses and deformations of glass panels, whether the vertical substructure element is a direct linear support of the panel or indirectly receives the concentrated force from point support to which the glass panel is directly attached. Avoiding vertical elements in the load bearing facade substructures in the case of a panel which has a height of at least twice its width significantly affects the increase in material consumption and the thickness of glass panels.

Analyses indicate that the dominant structural characteristics of facade glass panels are influenced by the supporting system and connection type-geometry rather than the exact material properties. An elaboration of supporting systems and connections offers the possibility for an architect to create attractive and interesting forms of structural glass facades. The supporting system can undergo large changes in shape and appearance by virtue of the combination of different connection types, thus influencing the structural characteristics of glass facade panels and facade appearance.

The general conclusion is that the supporting system of glass facade panels has a direct impact on global processes of glass facade design and can define:

- visual identity of the facade,
- grid of the structural glass facade envelope,
- increased transparency of the facade, and
- flexibility of the interior space.

In the sense of increased transparency of the facade, the linear connection by using embedded-invisible sections and dotted connections on the level of the floor structure is the most appropriate, which is demonstrated in this paper.

The paper indicates various approaches to the design of structural glass facades in terms of supporting systems and connections, pointing out the challenges of their construction.

## Acknowledgments

This work has resulted from research within the COST Action TU0905 – Structural Glass – Novel design methods and next generation products, 2010-2014, and national scientific projects "Spatial, Environmental, Energy and Social Aspects of Developing Settlements and Climate Change – Mutual Impacts" and "Development and implementation of scientific methods in design and construction of economic structural systems by application of new technologies", financed by the Ministry of Education, Science and Technological Development of the Republic of Serbia, 2011-2015.

## References

Krstic-Furundzic Aleksandra, Tatjana Kosic and Jefto Terzovic. 2012. "Architectural aspect of structural design of glass facades /Glass skin applications/." In *Challenging Glass 3: Proceedings of international conference on architectural and structural application of glass*, edited by Freek Bos, Christian Louter, Rob Nijsse and Fred Veer, 891-900. Amsterdam: IOS Press BV.

Krstic-Furundzic Aleksandra, Tatjana Kosic and Jefto Terzovic. 2013. "Architectural aspect of structural glass roof design." In *Proceedings of COST Action TU0905 Mid-term conference on structural glass*, edited by Jan Belis, Christian Louter and Danijel Mocibob, 45-52. Leiden: CRC Press/Balkema.

Radimpex software. 2014. "About Tower 7." Accessed October 5. http://www.radimpex.rs/about.php?lang=en&id=1

Wurm, Jan. 2007. *Glass structures – design and construction of self-supporting skins*. Basel: Birkhauser Verlag AG.

# CHAPTER EIGHTEEN

# THE INFLUENCE OF NEW TECHNOLOGIES ON MODERN CITY FACADES

JASNA ČIKIĆ TOVAROVIĆ[1], JELENA IVANOVIĆ ŠEKULARAC[2] AND NENAD ŠEKULARAC[3]

**Abstract** | Science and technology development in the media brings changes in everyday life and affects our culture, art and architecture. Dynamic digital technologies are included in all segments of modern society and architecture. Architecture forms a part of the interactive environment of modern cities which are based on computer-technology models. Interactive design in architecture, especially the interactive designing of facades, has caused permanent changes in the conditions of architectural structure perception, as well as in the conditions of social communication and interaction within the urban environment. Nowadays, the un-changeability of the characteristics of facades belongs to history. Facades have become visually changeable and more dynamic. Their design meets a wide variety of their users' requirements and wishes. This paper aims to define the way these modern facades affect the appearance of new urban space and new communication forms and also to present the façade as a medium for specific social needs in urban environments. This paper deals with the role of modern technologies, interactive programming, the Internet and social networking in architecture and modern facade design.

[1] Faculty of Architecture University of Belgrade, Belgrade, Bulevar kralja Aleksandra 73/II, Serbia, e-mail: cikic.tovarovic@gmail.com

[2] Faculty of Architecture University of Belgrade, Belgrade, Bulevar kralja Aleksandra 73/II, Serbia, e-mail: jelenais@orion.rs

[3] Faculty of Architecture University of Belgrade, Belgrade, Bulevar kralja Aleksandra 73/II, Serbia, e-mail: nseki@orion.rs

**Keywords** | digital technologies, facades, interactive design

## Introduction

Over the last few years two important processes have affected all spheres of society and architecture – digitalization and globalization. As a result, cultural patterns in modern cities, technological possibilities in the field of architecture, urban perception and architectural perception have changed.

Post-industrial society is marked by: new parameters of space and time – real/virtual[4]; new social interactions – real time/entropy; and new materials and technologies in the field of construction–"smart materials"/digital and nano technologies.

The application of innovative technologies provides continual changes to physical space. According to some theories, duplicate reality has become one of the basic possibilities of a new approach in the field of architecture. The current image of architecture contains the integration of virtual and immaterial space with real, material space, where media-architecture causes an architectural object to connect these two realities.

## Architecture and digital technologies

Modern digital technologies have imposed significant changes upon architecture, design and architectural function. Above all, these technologies bring some important changes in the field of architecture by implementing intelligent and adaptable systems, where architectural space can adapt its parameters and connect them directly with users, passers-by and beholders. The main potential of information technologies is seen in creating emotional and personalized aspects of urban and architectural space (Puglisi, 1999: 7-8)[5]. According to architect Tojo Ito "the main objective of architecture is to connect us with the digital environment".

Also, computer-based digital media enable the process of experience transposition from other art forms into architecture as well. Changeability and dynamism are new forms of architectural expression. Yet, there is a

---

[4] Šuvaković M, p. 661. Virtual means a specific form of connecting technical systems with human beings. "Based on the post-modern conception, virtual reality is not the effect of individual surface (image, display) but a global or ambient simulation of reality."

[5] "Three main characteristics of a new architectural (hyper-architectural) media space: immateriality, sensitivity and multiplemedia."

real risk of putting up with the monotony of commercialized society in the field of modern city architecture.

The works of several authors reveal that some theorists (Baudrillard, Gibson) are critical of digital informational environments, explaining that media and technologies create a world of illusions, where consumers values are dominant and media ideologies are defined as "seductive cyberspace technologies".[6]

## Architecture and the media environment

Urban displays, media facades and media architecture are the main forms of media intervention in urban environments and they represent a unified media environment (Tscherteu, 2011). Urban displays are subsequently applied to architectural structures in urban environments. Their construction, form and content exist independently of architectural buildings. A media facade is a hybrid facade in both functional and aesthetic senses and represents the integration of architecture, technology and content (Haeusler, 2009: 13-14). From a technological point of view, a media facade is a structure that includes artificial lighting and/or kinetic elements that are able to provide dynamism and changeability as one of the major requirements for their proper functioning. According to Tscherteu, media architecture implies an interactive dimension to media facades, considering "the possibility of architectural space to gradually adapt along with an active role of a subject".

## Interactive architecture

"A particular system can be defined as an interactive system if it is flexible enough to change and cause another system to change as well (when connected to that system), through the connection made between them" (Haeusler, 2009: 230).

When trying to analyse the concept of interactive architecture, some of the precursors to interactive art[7] should be taken into consideration. When it

---

[6] Baudrillard, J., *Simulation and Simulacra*. Fiction novelist William Gibson coined the term cyberspace in his novel Neuromancer (1984); Gibson defines the term as "the space where the information found in virtual environment can be vizualized as geometry structures" (See Šuvaković M. p. 125)

[7] Šuvaković, M. p. 279. Interactive art work is a form of art that involves "network performance or the situation where there is a hardware and software feedback

comes to historical examples of architecture, ones that are not directly related to digital media, one should mention Le Corbusier and his Philips Pavilion designed for the Brussels World's Fair in 1958. The Pavilion was the project where Le Corbusier expanded his ideas on design by adding music and visual media, thus changing the main language of architecture through its connection to other art expressions.

How did the concept of interactivity gain such an important role in current architectural research? There are a few reasons. Firstly, interactivity includes modern communication systems based on making choices and creating hypertext[8] systems based on digital media. Secondly, interactivity is focused on a subject (changeability of architecture, reshaping and personalization) instead of an object which has "an absolutely unchangeable nature" (standardization, duplication, serial production). Thirdly, in the context of time, "interactivity is a concept of continual spatial reconfiguration, changing the boundaries of time and space that used to be regarded as unchangeable". Lastly, interactivity includes the main characteristics of computer systems: that is, the ability to create interconnected, changeable "sets of information that can be continuously changed" (Saggio, 2005: 23).

Urban public interactivity is realized through the movements of numerous different people; the very design of these areas directs the type of activity taking place there. Media architecture creates specific social interactions in a particular location; on the other hand, they are also the result of certain activities taking place within the location (Tscherteu, 2011).

Interactive media architecture can facilitate the process of creating modern cities' identities, since the inhabitants themselves should take part in creating these identities through their urban space activities (fig. 19-1). Some authors have a dilemma and wonder if media facades should decontextualize their surroundings or accentuate the cultural identity (Singer, 2010: 54).

---

between computer systems and recipients, or in other words – there is a recipient intervention through technology systems".

[8] A prefix hyperis a Greek word υπερ (huper) that means over, above. Hypertext is writing a text using various possibilities (computer links) in order to offer a certain form of interactivity and enable indirect connection with additional information apart from the text and different access orders.

*Figure19-1. PSD-Bank, Munster, Germany, 2008, arch. Kronhagel. Creation of city identity: Innovation and Change*

## Interactive facades

The application of interactive programming models causes media facades to become a part of interactive architecture. A system of interactively programmed media facades is an intelligent system that provides specific responses after the users have applied specific sets of inputs[9]. Similarly to emotional behaviour, responses are not always predictable. In most cases, the system is programmed to provide more responses to input sets. Facade "responses" are in the form of sounds, light or projection (text, graphics, photo, film, music etc.) (figs 19-2 and 19-3).

The interactive programming model implies that "people are able to take part in architecture without being passive and dealing with static sets of conditions, but proactive individuals shaping both space and architecture" (Kronenburg, 2007: 209). Not only does the concept of media architecture include facades, but also some interior parts of a building that are affected by both facade activities and the public space a building is surrounded by.

Media facades represent anintegral part of the urban environment and are available for a great number of users. Mobile phones (SMS, bluetooth), laptops or touchscreens can be used in order to show personal messages on such facades and achieve a high level of identification with architectural objects, which is considered to be an innovation in architecture. Cameras,

---

[9] See more details about intelligent systems. An intelligent facade (according to Annemie Wyckmans, Norwegian University of Science and Technology, Trondheim at Inteligent arh.pdf) is a facade able to adapt to its surroundings, considering perception, response and activities; this facade causes a building to deal with new situations and solve any problem caused by interaction with the external environment.

sensors and other intelligent systems provide proper system functioning (fig. 19-4).

***Figures 19-2 and 19-3***. *ARS Electronica center, Linz Project GuitARS- is a project promoting the interaction architecture/music, architecture/logic game (Rubik's cube)*

***Figure 19-4***. *Dancing in the Rain by Pfadfinderei + The Constitute, Connected Cultures, Sao Paulo 2013 (http://connectingcities.net)*

Interactivity enables individuals and groups to connect with physical space[10]. "The individual's presence is scaled to match the scale of city. ... This is how light in the public realm can transform the public realm itself".The changeable characteristic of glass as a material supports a new paradigm in architecture (Carpenter, 2010) (fig. 19-5).

[10] Damon Seeley, partner in *Electroland* (Los Angeles)

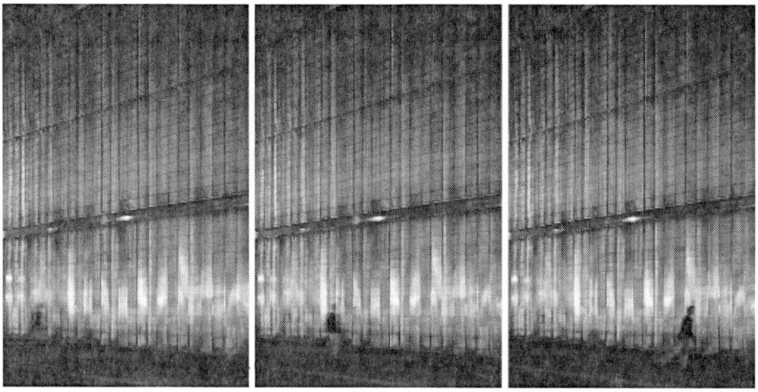

*Figure 19-5. Podium wall - 7 World trade center, has media content on both south and north glass facades and the system of video cameras recognizes pedestrians and their movements nearby the building itself and transforms them into bluish LED lines spreading over a seven-floor facade (Kinecity Llc, artist James Carpenter (http://kinecity.com/podium-wall-7-world-trade-center/)*

The use of a classic video game within an urban space and video animation is carried out through mobile phone applications where beholders are able to create their own image on a facade and e-mail it. One of the examples that should be pointed out is the interactive temporary facade placed on the already existing T2 tower of the National Library in Paris during the Nuit Blanche Art festival. This is a small resolution facade – the facade division system of facade segments is 20x26, 520 pixels on a surface of 3370 square metres on one of the most important buildings of Paris (fig. 19-6).

*Figure 19-6. Blinkenlights Arcade 2002, Paris, France (http://bilkenlights.net/arcade)*

A media facade can be a platform used by citizens who share their experiences. It is also a model for the interconnection of cities. The project *Connecting Cities* is European infrastructure of cities – they exchange their artistic and social experiences through media facades. The main objectives of this project are: networked cities based on media facades, thus creating a field of intercultural exchange; citizens becoming involved through their personal interventions, such as taking part in the urban public space; and artistic, social and intercultural processes being made available and sustainable within the urban environment. In his collection of essays “The Transparency of Evil” (1994:16), Baudrillard defines *transaesthetics* as a situation where art is everywhere but it is nowhere as a particular phenomenon.

## Conclusion

Over the last few years the concept of interactivity has become significant in the field of modern facade design. Media facades and their active light sources represent a suitable model for realizing this concept. The objective of architecture is to connect a subject to a digital environment through the process of creating events within urban public spaces. These events are able to change spatial activities, and to stimulate interaction among people and their re-integration into urban public spaces. As for the function, content and visual form, an interactive facade is the result of multiple social and cultural interactions and reactions.

## References

Bodrijar, Žan.1991. *Simulacija i simulakrum.* Novi Sad: Svetovi

Bodrijar, Žan.1994. *Prozirnost zla.* Novi Sad: Svetovi

Carpenter, J.(2010) “Light in public realm”. Accessed February 5, 2014. http://www.glassfiles.com/articles/light-public-realm

Haeusler, Hank. 2009. *Media facade*. Ludwigsburg: Avedition.

Luigi, Prestinenza Puglisi. 1999. *Hiperarchitecture Spaces in Electronic Age.* Basel: Birhauseer, The IT Revolution Series

Kronenburg, Robert. 2007. *Flexible: Architecture That Responds to Change.* London: Laurence King Publishing

Pokrajac, S. 2002. *Tehnologija i globalizacija*. Beograd: Grafolik

Rogina, K “Fragmentacija digitalnog oblikovanja”. Last modified February 28, 2006. http://www.a4a.info/articleview.asp?article_id=912.

Saggio, Antonino. 2005. “Interactivity at the centre of Avantgarde –architectural Research”, *AD- 4D space:Interactive Architecture*, vol. 75 no. 1. (Jan/Feb ): 5-11.

Schoch, Odilo. “My Building is my Display - omnipresent graphical output as hybrid communicators”. In *Communicating Space(s): 24th eCAADe Conference*

*Proceedings*, 610-616. eCAADe: Conferences. Volos, Greece: University of Thessaly, 2006.

Singer, Herald. 2010. *Mediatecture*. Vienna: Springer

Šuvaković, Miško. 2005. *Pojmovnik suvremene umetnosti.* Zagreb: Horetzki

Townsend, Anthony. 2004. "Digital mediated urban space: new lessons for design". Cambrige: *Praxis journal of writing+building.* 6:100-105.

Tscherteu, G.(2011)"Designing Urban Media Environmentsas Cultural Spaces" Accessed January 15, 2012.
http://www.largedisplaysinurbanlife.cpsc.ucalgary.ca/PDF/tscherteu_final.pdf

"Arcade". Accessed February 5, 2014. http://www.bilkenlights.net/arcade

"Connecting cities". Accessed February 5, 2014. http://www.connectingcities.net

"Image of architecture in electronic age-Architecture in Electronic Age is Figuration of Vortex of Information". Accessed January 31, 2014."
http://www.designboom.com/eng/interview/ito_statement.htm

"Media facades-Interactive brand communication in urban space". Accessed January 15, 2010.http://www.arclighting.de

# CHAPTER NINETEEN

# NEW COMPOSITE SLAB SYSTEM: LIGHTWEIGHT CONCRETE, STEEL SHEETING AND REINFORCEMENT

ZORAN ŠOBIĆ[1], JELENA MILOŠEVIĆ[2] AND MIODRAG NESTOROVIĆ[3]

**Abstract** | The construction industry is a vital part of the growth and success of a country. It is responsible for building the physical infrastructure that provides transportation and facilities for citizens, businesses, industries and institutions. Construction has a major influence on the economic wealth, societal well-being and sustainability of the built environment. This paper presents the structural behaviour of a new composite concrete slab system with profiled steel sheeting and lightweight concrete fill. The system has the potential to become well - accepted by the construction industry due to its many advantages over other types of floor systems. The slab is created by composite interaction between the concrete and steel deck with reinforcement to improve their bending resistant characteristics. The profiled steel deck performs two major functions that act as a permanent formwork during the concrete casting and also as tensile reinforcement after the concrete has hardened. The only additional nominal light mesh reinforcement bars that need to be provided are to take care of shrinkage and temperature, usually in the form of welded wire fabric. Steel possesses a unique material property unrivalled by other materials in that it can be recycled both up and down

---

[1] PhDc, Research Assistant, Faculty of Architecture, University of Belgrade, Bulevar Kralja Aleksandra 73/II, Belgrade, Serbia, zoran.sobic@gmail.com

[2] PhDc, Research Assistant, Faculty of Architecture, University of Belgrade, Bulevar Kralja Aleksandra 73/II, Belgrade, Serbia, jelena.z.milosevic@yahoo.com

[3] Professor, PhD, Faculty of Architecture, University of Belgrade, Bulevar Kralja Aleksandra 73/II, Belgrade, Serbia, enestorm@arh.bg.ac.rs

the product value chain without degrading its structural strength and other chemical properties. Open loop recycling allows an old structure to be melted down, and then, when it is recycled, it is re-melted to produce new appliances, structural beams, cladding and decking.

The presented research also focuses on the economic parameters of the topic, giving examples of some typical spans, in relationship with transportation needs, technology aspects and estimated time.

**Keywords** | Lightweight Concrete, Steel Sheeting, Reinforcement, Composite Slabs

## Introduction

Sustainable construction refers to the adoption of building designs, construction methods and materials that are environmentally friendly. It also means using materials and resources that have sustainable supplies and are readily available from many sources (Namdeo Adkui Hedaoo). Through sustainable construction, we will do our part to optimize the use of natural resources via recycling and the reuse of materials. This will also reduce our dependence on raw building materials, given the current disruption in the supply of concreting sand and granite.

The percentage share of the slabs in a structural system, in terms of weight, time, and cost performance is at about 50%. This means that the topic of optimizing slab systems is very important in the construction industry.

The maximum span length of a single span composite slab in contemporary use based on available steel deck floors in the market, is around 6m to 9m. The choice of systems is very common because these systems can save construction costs and time. If the span length can be increased by a factor of, for example, 1.5 or 2, significant cost savings can be expected from the elimination of some intermediate beams and their connections to girders. These potential advantages have motivated research in the area of composite slab systems. In this case, long span slab systems that do not cause any significant increase in the depth and weight of the slabs compared to regular span slabs are particularly attractive. This has been one of the main objectives of the study.

## Discourse of composites of steel, light and reinforced concrete

Steel is an excellent reusable material. Independent agencies (and some steel producers) around the world have performed life-cycle analyses on the environmental impacts of using steel. Based on the results, informed designers can confidently specify steel products in their various forms for use in projects of all sizes, from single storey and low rise to high rise buildings. Steel can be recycled repeatedly without any degradation in terms of its properties or performance quality.

Steel has one of the highest strength to weight ratios of any construction material. Steel framing can weigh only one-third as much as traditional construction materials. When connected by fasteners, a steel frame is stronger than traditional systems. This provides savings with regard to the foundation and the amount of carnage needed on site.

## New Composite Slab System

The presented composite slab (Figure 1) is a mixed construction system made of worked steel sheeting, into which the appropriate reinforcements and the spreader mesh are inserted in order to avoid cracking due to retraction and temperature effects. Afterwards, the concrete is spilled. In addition, concrete works jointly with steel, giving the advantage of both materials. Further, there are lightweight concrete fillings, with additional empty space, for MEP systems.

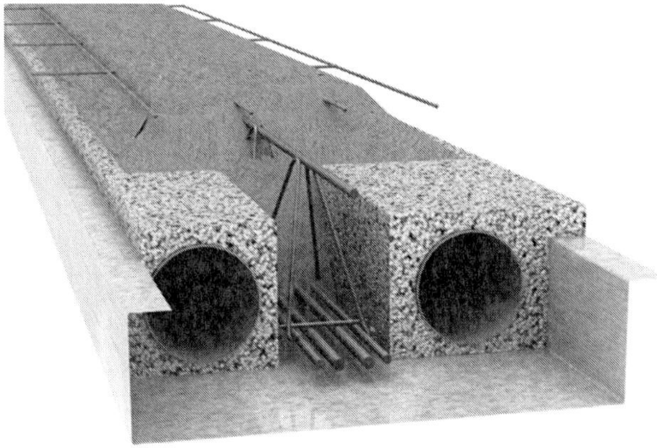

*Fig 1. Composite Slab System: Steel Sheeting, Lightweight Concrete Filling, Reinforced Concrete*

This system is specially designed to be installed over steel structures, though it can be used with concrete, wooden and masonry structures as well, provided that the fixing and flange conditions are the optimum ones. It is suitable for a wide range of applications, such as:

- Industrial buildings
- Offices
- Public buildings
- Warehouses
- Department stores
- Sport centers
- Houses
- Garages

The major functions of the subjected slab system are:

- Formwork
- Useful as a safe working platform
- Bracing structure
- Self-supporting weight and supports the overloading to which it is subjected
- Facilitates access to installations
- Acoustic and thermal insulation

The aimed for and earned advantages are:

- To lessen its own weight
- Avoid bracing
- More economic
- Facilities in stockpile
- Quick installation
- Does not pollute other materials
- Cleanliness
- Makes the movement of the building plants easier, since bracing is not necessary

In the current study, deep steel deck profiles are investigated (Figure 2). Two design phases have to be considered in the development of these new deck profiles for long span composite slab systems, namely the *construction* (non-composite) phase and *service* (composite) phase. The construction phase considers the strength and stiffness of the steel deck as a working platform that is subject to concrete self-weight and construction loads.

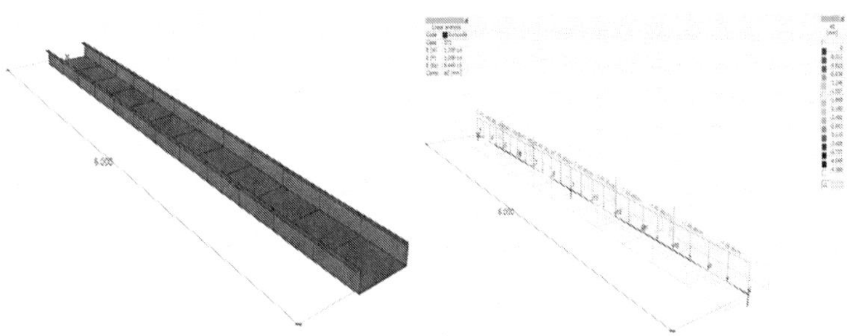

*Fig 2. Composite Slab System: Steel Sheeting, Lightweight Concrete Filling, Reinforced Concrete*

This phase is important in the determination of the required deck stiffness. In the following, it is shown that when a long span system is involved, the deflection (stiffness) limit state becomes very crucial (An L.). The service phase deals with a composite section of a steel deck-concrete slab that is subject to occupancy loads.

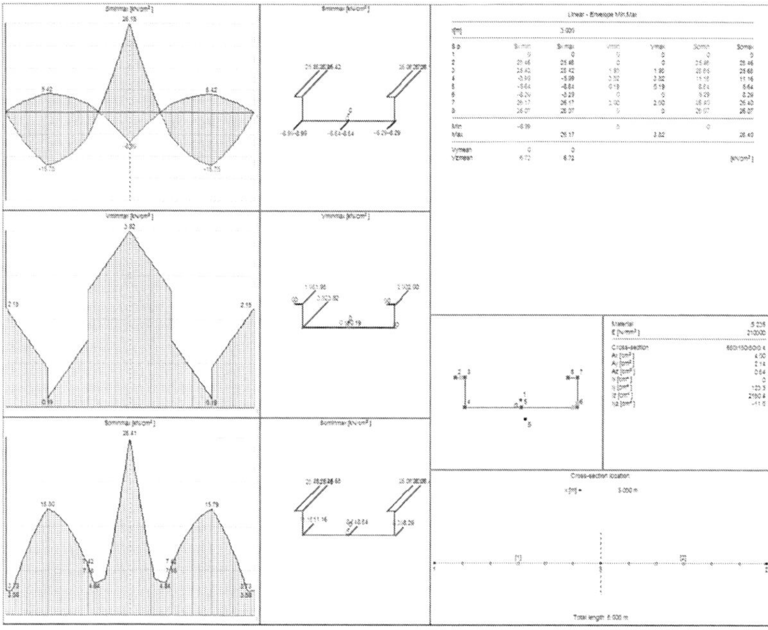

*Fig 3. Composite Slab System: Steel Sheeting, Lightweight Concrete Filling, Reinforced Concrete*

## Construction Phase

As previously mentioned, this design phase considers the strength and stiffness of a steel deck due to the fresh concrete weight. For typical span lengths, the flexural strength limit state is generally the governing condition in the design. For a longer span the governing condition is shifted toward the stiffness or deflection limit state. The steel deck weight is calculated based on the deck thickness that corresponds to the required moment of inertia for a certain span length with a given concrete self-weight plus the construction load.

## Exploitation Phase of the Assembly

In the service phase, predicted maximum loads of composite slabs can be calculated in various ways. The analysis was performed in the same way as for those with a typical span length.

## System elements

This paper focuses on a new slab system and presents the set of included elements. Basically, there are steel sheeting panels, profiled and designed for the construction and service phase; a lightweight concrete filling, lowering the weight of the package; and a reinforced concrete rib inside the composite slab system. Beside these, there are cardboard tubes, for additional lowering of self-weight, and for the MEP system positioned under and through the slab. The paper gives the scope of formwork elements, needed for production (Figure 4).

*Fig 4. Formwork and additional elements: a) Formwork, b) Lightweight fill, c) Prefabricated element, prepared for transportation and erection*

The system elements are designed to be lightweight and easy in terms of handling, erecting and positioning on the site, but they are also designed to occupy less space on the production site, in transportation and, of course, on the construction site. (Figure 5)

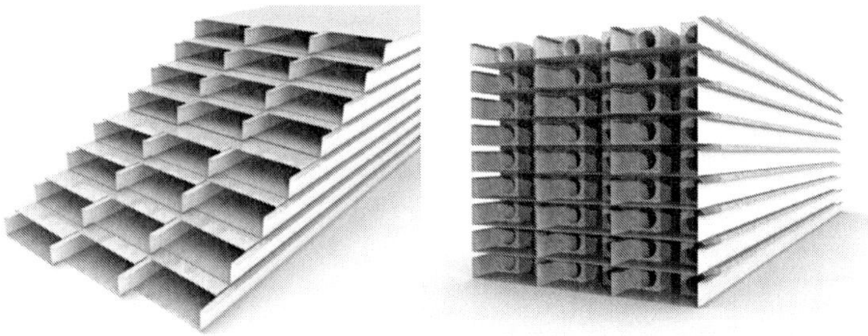

*Fig 5. Package of the parts; left: profiled steel sheeting storage; right: prefabricated slab elements storage*

Figure 6 shows the full range of elements, and Figure 7 shows the service stage of the composite slab with flooring. Lightweight concrete gives an additional insulating quality to the slab. Therefore, this system provides better conditions for floor heating systems in residential and public buildings.

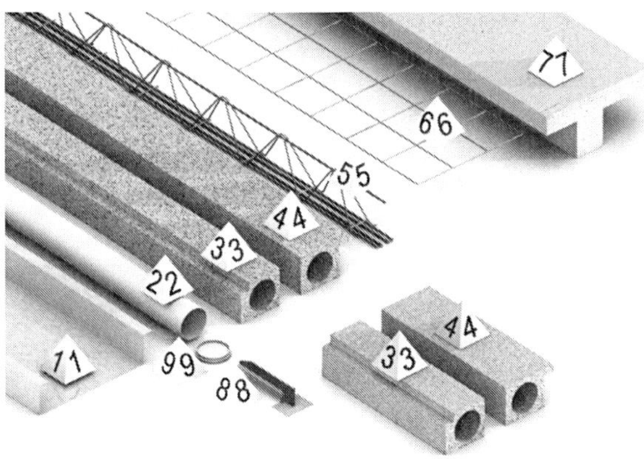


*Fig 6. Range of elements: 1) Profiled steel sheeting; 2) Cardboard tube; 3, 4) Lightweight concrete fil; 5) Reinforcement; 6) Steel mesh, 7) Bearing concrete fill; 8, 9) Additional accessories*

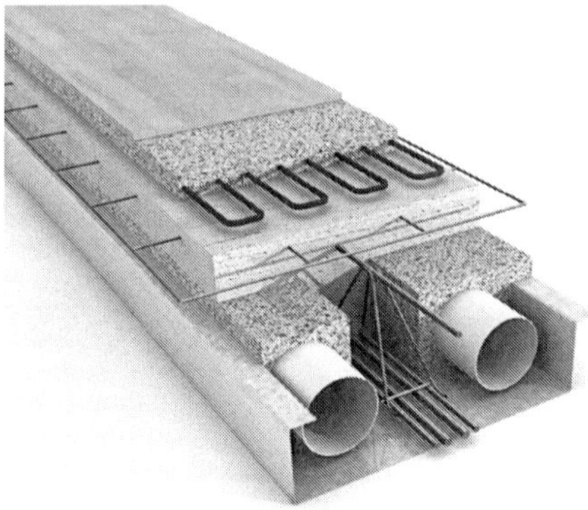

*Fig 7. Exploitation stage with floor heating and flooring*

## Conclusion and directions for further development

This paper has presented the new and innovative composite slab system which gives details about plenty of optimal preferences. More than simply this function, is it an opportunity for different materials to be used. One of the next goals is to develop the analogue system for wall partitions and the façade sandwich panelling system. Figure 8 shows the first images of this.

*Fig 8. Development of wall and façade panelling system*

## Acknowledgements

All of the material presented is part of the intellectual property of the corresponding author. Solution and all images, models and sketches by Zoran Šobić.

## References

An, L. 1993. "Load Bearing Capacity and Behaviour of Composite Slabs with Profiled Steel Sheet" Ph.D. Dissertation, *Chalmers University of Technology, Division of Concrete Structures,* Goteborg, Sweden (1993).

Namdeo Adkui Hedaoo. 2012. "Design of composite slabs with profiled steel decking: a comparison between experimental and analytical studies" *International Journal of Advanced Structural Engineering* (2012).

# PART 3:

# PLACES, TECHNOLOGIES AND RELATED FIELDS

# CHAPTER TWENTY

# THE ROLE OF CARTOGRAPHY IN MAKING A "SMART CITY": CASE STUDY OF INDJIJA

DRAGUTIN PROTIĆ[1], IVAN VUČETIĆ[2] AND IVAN NESTOROV[3]

**Abstract** | The goal of "smart city" electronic services is to significantly improve urban life and development with regard to different aspects. The services, built upon various Information and Communication Technologies, are usually designed to provide some sort of information to the users, and often the information has a clear spatial dimension. Modern digital cartographic technologies are exploited in order to model, analyse and bring spatial information to the users. The paper presents the case study of the city of Indjija in Serbia, where two EU funded projects, iSCOPE and eEnviPer, designed to establish different "smart city" services, have been implemented. eEnviPer is an e-Government service that supports the entire "environmental permitting" process. One of the elements within the service is Geographic Information System of the Environment which enables visualisation of spatial information related to the environment as well as some analytical capabilities. The i-SCOPE service is designed to provide information on solar potential assessment of the city roofs through an interactive 3D web map client. The technology is based on a 3D Urban Information Model. A number of user groups are expected to benefit from these services. eEnviPer platform is intended to improve the quality of environmental protection as well as to increase citizen participation in decision making, whereas i-SCOPE is intended to

---

[1] Faculty of civil-engineering, University of Belgrade, Belgrade, Serbia, protic@grf.bg.ac.rs
[2] Evrogeomatika, Belgrade, Serbia, vuceta87@yahoo.com
[3] Faculty of civil-engineering, University of Belgrade, Belgrade, Serbia, protic@grf.bg.ac.rs

influence the urban planning activities and increase renewable energy consumption.

**Keywords**| "smart city", web mapping, Urban Information Model, eEnviPer, iSCOPE

## Introduction

Introducing various web based ICT services in urban environments inevitably changes and improves different aspects of urban life. The services are generally designed to facilitate social processes and exchange of information between stakeholders as well as to produce new information, making a city "smarter". The information often has a clear spatial dimension, which results in an increasing demand for services that enable visualisation and analysis of spatial information.

In this paper, two projects are presented, both of which are aimed at installing ICT services for urban environments with integrated modern cartographic technology to provide spatial information. i-SCOPE service delivers information on solar potential assessment at building level. eEnviPer is an e-Government service that supports the entire "environmental permitting" procedure. Both projects are in the pilot testing and evaluation phase, with pilots being implemented in a number of European cities, including the Municipality of Indjija.

*Table 1. "Smart City" characteristics (source: Giffinger et al. 2007)*

| SMART ECONOMY (Compretitiveness) | SMART PEOPLE (Social and Human Capital) |
|---|---|
| Innovative spirit<br>Entrepreneurship<br>Economic Image and trademarks<br>Productivity<br>Flexibility of labour market<br>International embeddedness<br>Ability to transform | Level of qualification<br>Affinity to lifelong learning<br>Social and ethnic plurality<br>Flexibility<br>Creativity<br>Cosmopolitanism/Open-mindedness<br>Participation in public life |
| **SMART GOVERNANCE (Participation)** | **SMART MOBILITY (Transport and ICT)** |
| Participation in decision-making<br>Public and social services<br>Transparent governance<br>Political strategies and perspectives | Local accessibility<br>(Inter)-national accessibility<br>Availability of ICT infrastructure<br>Sustainable, innovative and safe transport systems |

| SMART ENVIRONMENT<br>(Natural resources) | SMART LIVING<br>(Quality of life) |
|---|---|
| Attractively of natural conditions<br>Pollution<br>Environmental protection<br>Sustainable resource management | Cultural facilities<br>Health conditions<br>Individual safety<br>Housing quality<br>Education facilities<br>Touristic attractivity<br>Social cohesion |

## What is a "Smart City"?

A "smart city" can be defined as a city well performing in a forward-looking way in six characteristics, namely Smart Economy, Smart People, Smart Governance, Smart Mobility, Smart Environment and Smart Living, built on the "smart" combination of endowments and activities of self-decisive, independent and aware citizens (Giffinger et al. 2007). For each of the characteristics, a set of factors can be defined (Table 1).

## i-SCOPE project

### General project information

i-SCOPE project is aimed at delivering an open source toolkit for 3D smart city services. One of the main objectives of i-SCOPE is to implement value added services in urban scenarios based on real users' needs and practices in three different domains:

- improved inclusion and personal mobility,
- energy dispersion and solar energy potential assessment,
- noise mapping and simulation.

The technology is being tested and evaluated in ten European pilot cities and regions. The pilot in Indjija will deliver a service for solar energy potential assessment.

The service allows users to define a temporal frame and to request the solar potential assessment information for the previously selected building(s) for a defined period of time. The information provided by the service includes:

- Electrical yield,
- Acquisition costs,
- Reduction of costs,

- Return of investment,
- $CO_2$ savings.

The favourable geographical position of Indjija in terms of solar irradiation and government incentives for solar energy exploitation (which are particularly attractive when PV panels are installed on roofs) make the "smart city" service potentially interesting for various target user groups:

- Local authorities (who would need the service for urban planning which is more oriented towards sustainable development and energy efficiency)
- Companies (that can use the information to develop their business)
- Citizens (who can learn about the potential of saving on energy bills or even make money from selling the electric energy produced by PV panels installed on their roofs).

## 3D Urban Information Model (UIM)

The i-SCOPE "smart city" services are built on 3D Urban Information Models (UIM). In particular, the service aimed at providing information on solar potential at building level requires 3D model of roofs; to calculate the best routes for mobility or visually impaired persons, information on path slope and barriers are needed, etc. It is clear that many other potential applications can be built upon 3D UIMs and that interoperability and standardization in creating such models are required.

In the i-SCOPE project, for building 3D UIM, cityGML, an OGC[4]standard, has been used. cityGML is defined as (cityGML 2012):

> *"..a common information model and XML-based encoding for the representation, storage, and exchange of virtual 3D city and landscape models. CityGML provides a standard model and mechanism for describing 3D objects with respect to their geometry, topology, semantics and appearance, and defines five different levels of detail. Included are also generalization hierarchies between thematic classes, aggregations, relations between objects, and spatial properties. CityGML is highly scalable and datasets can include different urban entities supporting the general trend toward modeling not only individual buildings but also whole sites, districts, cities, regions, and countries."*

Thematic information attached to buildings or their features (e.g. the amount of annual solar irradiation for a roof) is visualised and presented to

---

[4]Open Geospatial Consortium

the end-users through the i-SCOPE 3D web client (Fig.1). This is an example of how the cityGML is used for both information generation and its cartographic representation. 3D city models, like the one described, are predicted to play an increasingly important role in our daily lives and become an essential part of the modern city Spatial Data Infrastructure (Prandi et. al. 2013).

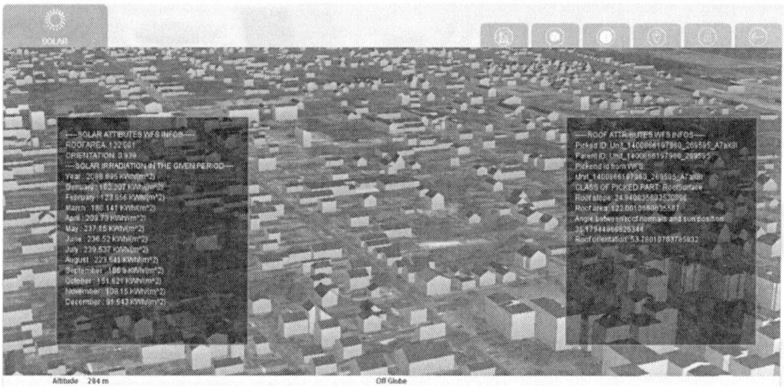

*Fig 2. The i-SCOPE 3D Urban Information Model of Indjija*

# eEnviPer project

## General project information

eEnviPer project, funded by the European Commission through the CIP-ICT-PSP program, aims to develop an e-Government service that will support the entire process of issuing approvals of Environmental Impact Assessment (EIA) studies. The service is built on a single multi-purpose cloud platform based on Service Oriented Architecture (SOA). It was tested in five different European countries, including Serbia. The Serbian pilot was run in the Municipality of Indjija. Building such a system is a complex task as there are many aspects to be considered and different components to be included. Use cases and workflows were created according to the national legislation and public authorities' practices. The system includes participatory Web 2.0 platforms to ensure easier and more effective public participation in environmental related decisions, Knowledge Management System (KMS) that provides relevant knowledge to the stakeholders and Geographical Information System (GIS) to enable access to geospatial environmental data and information and their analysis.

In Serbia, the Law on Environmental Impact Assessment (Official Gazette of the RS 135/04, 36/09) defines the EIA process including rights and liabilities of the stakeholders involved. Responsibility for managing the procedure and issuing environmental permits is divided between different levels of government. For small and medium size projects that are required to obtain consent on EIA study, local authorities (Municipalities) are in charge. The stakeholders involved and their actions within the process are: 1) Investor (or consultant on behalf of the investor): applying for permit, updating the case with new data (including the EIA study); Consultant: creating the EIA study for an Investor; Public authority: management of active cases, making data available for public inspection; Citizens: submitting complaints/comments on the active cases.

The Municipality of Indjija is one of the most developed municipalities in Serbia with a high level of investments, thus being an optimal pilot for the eEnviPer solution. Moreover, citizens in Indjija are accustomed to e-Government services since a number of them have been introduced over the past several years.

## GIS of the environment

Provision of relevant, accurate and up-to-date spatial data/information and the tools to measure, query and analyse it brings significant benefits to all stakeholders involved in the environmental permitting process. Problems related to these actions could seriously compromise the quality of the stakeholders' actions and decisions. eEnviPer project has recognized the significance of geospatial aspect of the EIA process and implemented an environmental GIS solution based on cloud computing.

The eEnviper platform uses the ESRI's ArcGIS Server for spatial data management and analysis. ArcGIS Server is used for creating and managing GIS Web services, applications, and data. Data and information layers present the other pillar of the GIS service. In the Serbian pilot, base-map (Fig. 2) consists of pseudo-natural coloured Landsat 7 image, roads and streets network, hidrography and forests and is aimed at providing general geographic information. Additional thematic layers (Fig. 3) can be selected by a user from the menu. They are collected from different sources and currently include:

- Spatial plan of the Municipality of Indjija (purpose and protection of the space);
- CORINE land cover for the year 2006;
- GMES HR soil sealing layer;

- Local registry of pollutants;
- Dumps and landfills.

The service also provides metadata for the geospatial data layers.

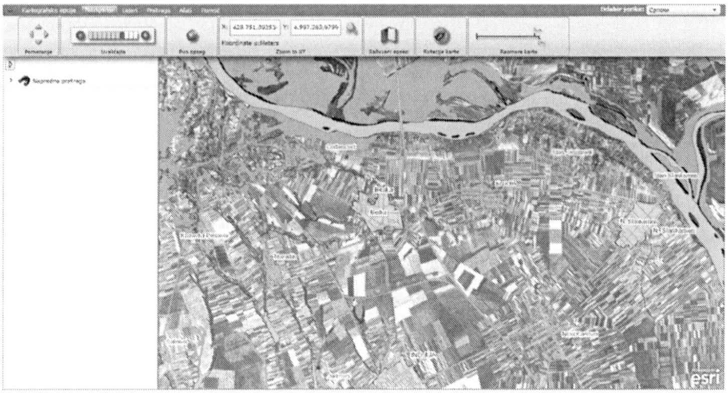

*Fig 2. eEnviPer base map*

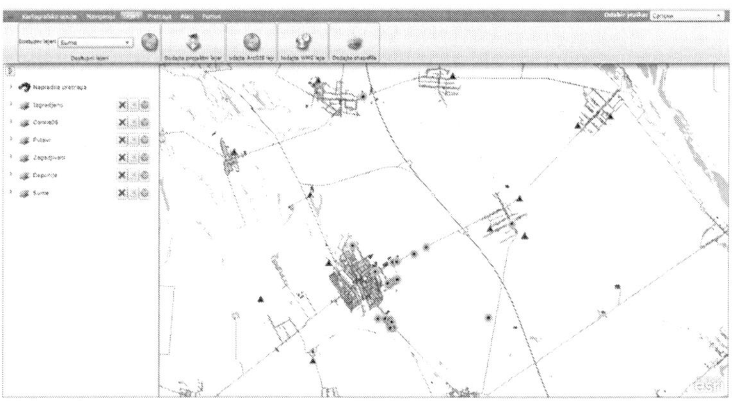

*Fig 3. eEnviPer additional data layers*

## Conclusion

"Smart city" services have the potential to significantly improve life in urban environments. The services are often based upon, or are using, spatial information to perform their work. For this reason visualisation of spatial information is clearly identified as a necessary part of many of the services. It opens up a new perspective for modern 2D/3D cartography, at the same time creating new challenges in terms of modelling, standards and functionalities.

## Acknowledgement

The paper represents the result of the research carried out on the project TR 36035, financed by the Ministry of Education and Science of the Republic of Serbia, and the projects eEnviPer and i-SCOPE that have been co-funded by the CIP-ICT-PSP as part of the Competitiveness and innovation Framework Programme by the European Community, contract number 297358 and 297284, respectively.

## References

cityGML. 2012. "Exchange and storage of Virtual 3D City Models." What is cityGML?. Last modified April 24, 2012. Accessed February 3, 2013. http://www.citygml.org/index.php?id=1523

Giffinger, R., Fertner, C., Kramar, H., Kalasek, R., Pichler-Milanovic, N., Meijers, E. 2007. *Smart Cities: Ranking of European medium-sized cities*. Vienna, Centre of Regional Science (SRF), Vienna University of Technology

Lombardi, P., Giordano, S., Farouh, H. and Wael, Y. 2011. "An analytic network model for Smart cities." in F. de Felice and A. Petrillo (eds.), *Proceedings of the 11th International Symposium on the AHP*, Sorrento, Naples, IT, June 15 – 18.

Lombardi, P. 2011. "New challenges in the evaluation of Smart Cities." *Network Industries Quarterly*. 13: 8-10

Prandi, F. et al. 2013. "Using CityGML to deploy Smart-City services for urban ecosystems."*Proceedings of UDMS2013*, London, May 29th-31st 2013

# CHAPTER TWENTY ONE

# RE-THINKING INFRASTRUCTURE PROJECT FOR METROPOLIS: LABORATORY GRANADA

JUAN LUIS RIVAS NAVARRO[1]
AND BELÉN BRAVO RODRÍGUEZ[2]

**Abstract** | Nowadays, urbanism must be able to reflect on the usefulness of certain forms of infrastructure, specifically upon transport infrastructure, when considering new policies for distribution of land uses. In the context of a new culture of sustainability and urban efficiency, this kind of urbanism seeks a reduction to the demand for mobility and has accordingly sought to design solutions.

The coexistence of nature and artifice in metropolitan contexts is currently an important issue on which to re-think and re-know its many possibilities. We could consider "place" as "urban and territorial material", re-interpreting the quality of places. We could also look at it through the classical argument of "points + lines + spots" which is inherited from landscape ecology. These and other methods are useful for understanding the urban fabric. They provide new meaning to the fabrics' intrinsic values, thanks to the presence of infrastructure.

From the perspective of the transport infrastructure project, this reflection should facilitate a methodological evolution, aimed at creating a new technology that understands design as a negotiation between the demands of citizens, and the spatial offer of "place". This technology would overcome the theoretical distance that exists between the perspective of the city as an autonomous compact entity, and the "splintered" present day

---

[1] University of Granada, Laboratory of Urban Planning, C/ Severo Ochoa s/n, E.T.S.I.C.C.P. Planta -1, 18071, Granada (Spain), juanluisrivas@ugr.es
[2] University of Granada, Laboratory of Urban Planning, C/ Severo Ochoa s/n, E.T.S.I.C.C.P. Planta -1, 18071, Granada (Spain), bbravo@ugr.es

metropolitan reality, defined by more open lexical and combinatory systems that are overlapping with conflict.

Thus we see, now more than ever, an extensions of the mechanisms of creative projects; the transforming of some spatial concepts into instruments such as outline, crossing or section, fusing landscapes, hybridizing functions, or of adding different times of the territory, etc., are also technological keys, for the integration of large scale infrastructure.

**Keywords** | Urbanism, Infrastructure, Design, Metropolitan

## Introduction

Urban design always attempts to recompose complex entities through an analysis of specific elements. It frequently uses a constructive methodology that observes parts or components of the whole, and serves to evaluate protagonism as well as to generate hierarchies. Using such fundamental criteria, an idea of the object of study or intervention is formed, irrespective of whether it is a specific area, a neighbourhood, a street or a complete city.

> *"Constructing culture from a wide base, with a critical capacity to select the past and a long range lens when selecting criteria of exploitation and utility (Solá-Morales, 1981)"*

A few years earlier, the architect Carlos Aymonino explained that in order to understand the significance of a city, it is necessary to identify its physical range, analyzing its morphology while classifying the elements, and understanding also the possible relationships between them.

> *"The construction and growing specialization of different productive activities, the integration, constantly variant, with a power of more or less expansion, constitute the economic-social causes of those effects that directly determine the quantitative characteristics that define cities: that is, the coexistence of spaces with a certain grade of separation, used at distinct moments by the same number of people. (Aymonino, 1981)"*

In Aymonino, the importance given not only to identifying the elements, but also to their spatial and temporal intertwining, can be seen. At the same time, he evaluates the constant process of transformation as well as that of the movement of the elements that give reference to a place, and in doing so not only forming a technical or formal point of view, but also through internal organization, which is characterized by a multiplicity of hierarchical relationships, that are independent and intensely selected.

## The role of infrastructure in the metropolitan context

Among contemporary landscapes and daily movements, a drifting theoretical space can be observed, incapable of moving in the correct direction, lost in the attempt to understand what is going on, and infinitely slower than an urban reality which is years ahead of our own, in a permanent state of mutation.

The dominant concern in the majority of Urban Planning forums is the search for meaning in urban events and the consequential responsibility of architects, urban planners and other increasingly diverse and qualified professionals in the territorial construction field to create such meaning.

## The multi-scale condition

The multi-scale condition of the urban events is evident in all places. All urban "objects" possess this capacity for creating relationships at a short distance, a medium-distance and at a great distance (Rivas, 2006).

Mobility infrastructure establishes an ideal framework to fragment disciplinary borders, reinforce theoretical compartments, techniques and proposals which are created out of sectorality; in this way, they are able to respond to a fortunately ambiguous reality.

## Infrastructure for the mix

Territorial lines, distribution and access channels, as well as intercity motorways, are even more intense and permanent vectors in our daily lives. By making them the object of study, it is possible to understand, perhaps in a more palpable manner than with different forms or uses (public spaces, natural spaces, logistical centres, residential packets, etc.), the potential of a priori labor which is hybrid, inter-disciplinary, complex, and which offers grand spatial capacity.

Among our landscapes and contemporary movements, the need to reconcile a structural view of territorial regulation with another sensible one is shown, in light of the minimal events which occur in the area.

The multiple speeds superimposed in any roadway which crosses the city or its area, and the capacity to establish smaller or greater relationships than its size, position as well as the continuity of these offers, makes it more and more necessary for the project to respond to this multi-scaled nature.

## Re-thinking infrastructure project

Contemporary urbanism must be able to reflect upon the usefulness of certain infrastructure, transport infrastructure mainly, and be based on new policies for the distribution of land uses. In the context of a new culture of sustainability and urban efficiency, this kind of urbanism seeks a reduction on the demand for mobility and expressly designed solutions.

The coexistence of nature and artifice in metropolitan contexts is currently an important issue with many possibilities that need to be re-thought and re-known. We could consider "place" as "urban and territorial material", according to the way Viganò (1999), Solá-Morales (2008), Zardini (2005), etc., teach us from their different scales, through re-interpreting the quality of places.

From the perspective of the transport infrastructure project, this reflection should facilitate a methodological evolution aimed at creating a new technology that understands design as a negotiation between public demands and the spatial offer of "place". This technology would overcome the theoretical distance that exists between the perspective of the city as an autonomous compact entity, and the "splintered" present day metropolitan reality, defined by more open lexical and combinatory systems, overlapping with conflict.

*Fig. 2-1: Aerial view of Jardins Wilson Project in Saint Denis, Paris, by Michel Corajoud (Lotus 110, 2001)*

*Fig. 2-2: Urban section of Jardins Wilson Project in Saint Denis, Paris, by Michel Corajoud (Lotus 110, 2001)*

## Space through its section: widths and links

Let's talk about urban planning projects in which the infrastructure is wisely domesticated without underestimating it, without effacing it, but instead finding the formula which will be woven into the fabric, incorporating it into the landscape of the city as a "positive" element (Smets, 2001).

From the simplicity of this gesture, the Jardins Wilson Project in Saint Denis de Paris, by Michel Corajoud, is an approach to the work with sections of main streets which cross a consolidated urban fabric, through a neighbourhood in transformation, on the outskirts of Paris, from its industrial vocation to an increasingly residential purpose.

By increasing the surface level of the corridor occupied by the infrastructure, a great quantity of new areas can be achieved, which can be occupied by the required public space. Together with this idea of a "complex section", it is very important to assess the strength and intelligence of the project as it obtains some excellent dimensions for the functioning of space from this process of sinking traffic, and while it necessitates "interesting" lengths and widths in the dialogue between one facade and the other, in order to allow for tree-filled spaces of great size.

## The moment of the crossing: possibilities and events

These projects have another complexity increasing mechanism in common, namely the interaction of the horizontal diversity. Now we consider the rehabilitation of the processes which multiply the horizontal channel; in other words, rehabilitating segregation and differentiation in the connected routes, roads, and different traffic.

Why is this moment of crossing extraordinary? As the possibilities for differentiated movement are generated, the number of directions which must be organised increases, because the crossing and the consumption of space in the infrastructure is much greater and does not take advantage of this spatial capacity in areas which are exclusively dedicated to traffic. These are hyper-connected places which are references for multiple roads, with very high agglutinating potential. If we play with the separation of the different channels instead of their concentration, we obtain accessible "air", which at the same time possesses an unprejudiced aesthetic sense, enabling experimentation with its form and function.

In this way, for example, the MVRDV project for the Stadsboulevard a 20 in The Hague intersperses residences, public space, offices, etc. within the motorway's environment of freedom and openness.

## The surrounding area of the infrastructure: implied space and architecture

The edges of infrastructure should be observed in order to interpret how spaces of conflict (in friction with the road) can be recovered and offered to all of society in the form of natural space, urban space or cultural space. In this way, what is in question is the very environment of the infrastructural network, its very shape as well as the forcefulness of its placement.

There is an architecture which is attached to the machine of mobility. This architecture emerges surrounded by a logic of noise and velocity, somewhere between its physical and environmental ends, in the tight strip which separates it from the urban or the natural; essentially, a border space between the two mediums. The space planning close to the motorway has to find arguments for use and scale which force these 'nowhere lands' to understand the means by which infrastructure bursts in.

## Infrastructural typologies

To be able to present synthetically, a "reconsidered" infrastructure in the previously described terms, the means by which these objects are implanted in the territory should be distinguished. These means can be seen as "typological infrastructure", permitting an understanding of the specific needs and classifying the competencies. At the same time, they are useful in order to understand the multiple sizes and scales of relationships that can unite into a single infrastructure. These are superimposed technologies in a single infrastructure that try to respond, in many cases, to the demands of the distinct, congregated typologies.

There are three modes or types of infrastructures in this sense:

a) Nodal infrastructure or the nodal situations of the infrastructure, such as roundabouts, various nodes, bridges, stations, parking lots etc.
b) Linear infrastructure or those that are recognized for their linear condition, such as highways, shorelines,. suburban roads, arterial streets, etc.
c) Infrastructure related with an area that constitutes systems or more complex groups. These are networks or elements that are a sum of infrastructure and that branch out into the recognizable landscape or territory: suburban spaces, airport surroundings, industrial areas, territorial links, etc.

## Metropolitan project: laboratory Granada

The following are examples of infrastructure projects that would constitute a new way of dealing with infrastructure, as a territorial or urban object. At the same time, infrastructure is treated as an element that belongs to a landscape and is integrated in its forms and functions. This kind of project, or infrastructural urban surroundings, observed and worked on in the Urban Design Laboratory at the University of Granada, explains the characteristics or qualities necessary for the achievement of the two fundamental objectives:

_ First, they peruse the integration of the scales and urban time as previously described.
_ Second, they try to reinforce the identity of the new metropolitan entities, seeking to increase cohesion between the pieces of the new urban systems.

The growth of southern Granada can only be understood in the context of its defining geographic conditions. The presence of the Genil River totally changes the inertia of the linear forms, deforming the north-south directional axis and approximating the points of origin of the metropolitan relationships to the two historic bridges that cross it (Puente Blanco and Puente Verde). This deformation creates a network of radial and circular lines where a rectangular, isotropic network would normally exist. In turn, it develops very special metropolitan relationships of proximity and distance between the city and the exurban settlements.

*Fig. 4-1: Aerial view of metropolitan south of Granada.*

The radial form is emphasized by suburban roads with extremities clearly defined by the nucleus that the Communicate (Huétor vega, Monachil, la Zubia, Dílar, Ogíjares, Armilla) and Granada as a common destination and point of exchange between them. These roads play a predominant role as a structuring element in the interior growth of the city to the south and in the metropolitan form of this southern area.

Thus, for example, urban infrastructure is situated on the edge of these suburban roads, where they are more accessible and visible. Each line acquires a character from the point of view of the collective spaces and residential neighborhoods or local networks.

*Fig. 4-2: Connection between Granada and Huetor Vega.*

*Fig. 4-3: Structure of suburban roads. (Bravo, Belén. Doctoral thesis in progress)*

## The project on suburban roads

Opposite to this traditional connectivity, the new approaches of metropolitan politics impose the use of new roads that reconfigure the territorial accessibility map without taking into account how this might affect the meaning of the urban uses and activities. The projects on these suburban roads are both necessary and new revitalizing actions for these weak infrastructures in order to make them compatible with contemporary needs and to modernize the traditional section. But it is probably much more about reconfiguring the surface; the way that Joan Busquets (2006) explains it in his projects, namely that the qualities of integrated landscape, multimodal capacities and the aggregative capacity in relation to the different traffics and requirements of metropolitan movement can be maintained,.

## Conclusions

Beyond the traffic's infrastructure, the urban or natural infrastructures must generate guidelines for the colonisation of its necessary uses on its near edges, or incorporate apparently strange uses into these in the surrounding environment, in order to allow for their absorption as participating elements of the whole.

These are some of the challenges of the infrastructural projects in times of urgent necessities to economize resources, a technical instrument that should be at the service of the reutilization of the pre-existing objects, giving them new meaning and avoiding unnecessary infrastructure that starts from scratch when it comes to their relationship with the metropolitan forms and the territorial identities in construction.

## References

Aymonino, Carlo. *El significado de las Ciudades*. Madrid: Blume, 1981.

Bravo, Belén. "La ciudad doméstica". In *III Seminario de Investigación en Urbanismo*. (Barcelona: DUOT, Universidad Politécnica de Catalunya, 2010), 45-55.

Casariego, Joaquin, and Guerra, Elsa. *Flowpolis: la forma del espacio nodal / Flowpolis: nodal the form of nodal space*, EU (Especiales de Urbanismo) nº4. Las Palmas de Gran Canaria: Ed. Ayuntamiento de Las Palmas de Gran Canaria, 2008.

Busquets, Joan. *Ciudades X formas: una nueva mirada hacia el proyecto urbanístico = cities x lines: a new lens for the urbanistic project*. Rovereto: Nicolodi, 2006.

Gómez, J.L.; Rivas, J.L.; Cabrera, D.; Reinoso, R. *Un Curso de urbanismo: 2002-2003: proyectos urbanos y territoriales: aglomeraciones de Granada, Sector Sur*. Granada: Universidad de Granada, 2006.

Maas, Winy; Van Rijs, Jacob and De Vries, Nathalie. "Vivre au bord de l'autoroute". In *Quaderns d'arquitectura i urbanisme*, no. 218 (1997): 54-61.

Portas, Nuno; Domingues, Alvaro and Cabral, Joao. *Politicas Urbanas. Tendências, estratégias e oportunidades*. Lisboa: Fundação Calouste Goulbenkian, 2004.

Smets, Marcel. "The Contemporary Landscape of Europe's Infrastructure". In *Revista Lotus Internacional*, Cittá globali, no. 110 (2001): 110-115.

Solá-Morales, Manuel de. *De cosas urbanas*. Barcelona: Gustavo Gili, 2008.

Solá-Morales, Manuel de. "La identidad del Territorio". In *Quaderns d'arquitectura i urbanisme*, no.1 (1981): 3.

Venturi, Robert; Scott Brown, Denise and Izenour, Steven. *Learning from Las Vegas.* London: The MIT Press, 1972.

Zardini, Mirko. *Sense of the city: an alternate approach to urbanism*. Quebec: Canadian Centre for Architecture, 2005.

# CHAPTER TWENTY TWO

# PERSONAL RAPID TRANSIT: A SUSTAINABLE URBAN TRANSPORT SYSTEM

LJUPKO ŠIMUNOVIĆ[1], LUKA NOVAČKO[2] AND MARIO ĆOSIĆ[3]

**Abstract** | High urbanisation processes and population growth in cities, together with the desire of city residents to own a car, has led to the present situation in which huge traffic-related problems occur. The rapid growth in the number of private cars has not been followed by an appropriate supply of road infrastructure. Energy efficiency has also dropped behind the number of private cars in the world. This has produced problems such as congestion and excessive energy consumption – both of which are serious threats to the urban environment. Therefore, urban transport has entered a so-called "vicious circle", in which the increase of private car usage leads to reduced travel speeds and increased travel times for each person in traffic. There have been many efforts to change people's travel habits and to encourage the use of other modes of transport. Urban transport planners recommend cycling, walking and public transport as alternatives to private cars. The greatest disadvantages of walking and cycling are the difficulty in travelling longer distances, the influence of weather conditions and issues with luggage. From the traffic participant's point of view, public transport lacks flexibility in terms of both travel times (fixed timetable) and travel routes (pre-defined routes of public transport). However, walking, cycling and traditional mass transit do not fully comply with the mobility requirements in progressively

[1] Faculty of Transport and Traffic Sciences, Vukelićeva 4, 10000 Zagreb, Croatia, ljsimunovic@fpz.hr
[2] Faculty of Transport and Traffic Sciences, Vukelićeva 4, 10000 Zagreb, Croatia, lnovacko@fpz.hr
[3] Faculty of Transport and Traffic Sciences, Vukelićeva 4, 10000 Zagreb, Croatia, mcosic@fpz.hr

growing cities. Solutions to the problems mentioned above are sought by introducing new technologies. Personal Rapid Transit (PRT) is a proposed solution to the transport problems of the future, providing many benefits not only for passengers (safety, speed, privacy, comfort and low transport costs), but for entire cities (low introduction costs, low operation costs, low noise production and a negligible impact on the environment). The emphasis of this article is given to the description of PRT systems, and not to their technical details.

**Keywords |** Urban Transport, PRT, Sustainability, Mobility, Congestion

## Introduction

The private car is still the dominant mode of transport in urban areas because of its advantages. It allows people to travel "from door to door", giving them more freedom and comfort when compared to other modes of transport. Travel time is much shorter by private car if compared to public transport. Therefore, cars as a mode of transport are considered to have a good image in society. However, the increase in the number of private cars in combination with limited urban space creates traffic jams which result in higher noise and emissions, which have become threats to the urban environment. The cost of the time lost in traffic jams is estimated to range between 0.5% and 2% of GDP in European countries. In 2010, these costs were expected to double. The problems are trying to get solved by applying the design of new vehicles with the use of old technology, but people still use private cars. In more recent times, there has been a great interest in a new and sustainable transport solution named Personal Rapid Transit (PRT). Personal Rapid Transit is not just a new vehicle – it is an entirely new form of transportation system which integrates road transport technology (from private cars), navigation technology, accuracy and communication (from air transport), and finally, security and infrastructure (from rail transport) (Mogge, 2009). This is almost the only system which can compete with private cars because of its advantages. The characteristics of PRT systems are the following: the vehicles are almost instantly available, so there is no actual waiting time like for private cars or taxis (because of the non-stop driving resulting in high-frequency arrivals); it does not require taxpayer subsidies and has no impact on the environment. However, the experts are not in agreement in their opinions regarding PRT: some consider that PRT has a great potential to solve traffic problems, while others believe that PRT will never become operational and that it slows down the development of light rail systems. But the actual truth lies somewhere in the middle.

## Personal Rapid Transit Definition

The Advanced Transit Association (ATRA) (Advanced Transit Association – ATRA, 2003) has provided the following definitions of Personal Rapid Transit:

1. Direct source-to-destination service with no need to transfer or stop at intermediate stations.
2. Small vehicles available for the exclusive use of an individual or small group travelling together by choice.
3. Service available on demand by the user rather than on fixed schedules.
4. Fully automated vehicles (no human drivers) which can be available for use 24 hours a day, 7 days a week.
5. Vehicles captive to a guideway reserved for their exclusive use.
6. Small (narrow and light) guideways are usually elevated but can also lie at ground level, near ground level or underground.
7. Vehicles able to use all guideways and stations on a fully connected PRT network.

Personal Rapid Transit is an alternative means of transport within dedicated areas or as a feeder service to existing (public) transportation systems. Public Rapid Transit has the following fields of application: airports, business parks, campuses, city centres, new and eco towns, resorts, national parks, and entertainment parks, etc. Over long distances, PRT systems cannot compete with the private car for the time being. However, a modal shift can be achieved if PRT is connected into the intermodal chain with well-organised public transport. In order to shift passengers onto public transport, they have to see the benefits of such a transport system, which is accomplished by ensuring that the entire system has no weak links whatsoever. Feeding a train system with buses does not work well, so passengers are reluctant to use two services if the waiting time is unpredictable. This problem can be solved by integrating PRT collector services into the rail system, as shown in Fig. 1.

The U.S. Department of Transportation (Irving & Corporation, 1978) suggested a more generic term for Shuttle Loop Transit (SLT), Group Rapid Transit (GRT), and Personal Rapid Transit (PRT) – Automated Guideway Transit (AGT). Automated Guideway Transit is defined as any kind of a transit system which incorporates automated vehicles moving on a fixed path with exclusive priority. The path may be positioned underground, at ground level of above ground, but in any case, it is separated from streets and pedestrians, so that pedestrians and vehicles do not interfere with the movement of automated vehicles.

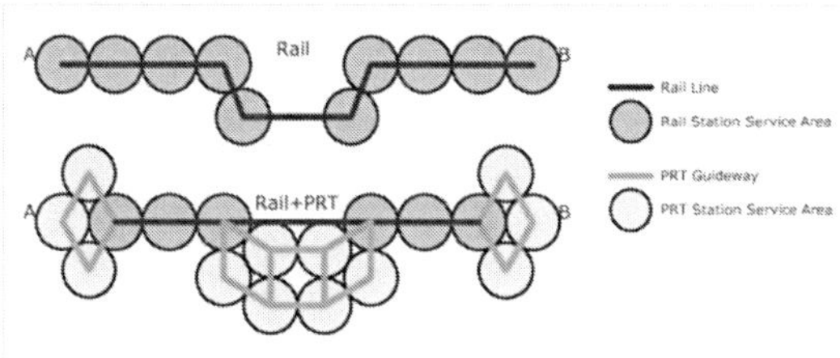


*Fig. 1. Comparison of common rail system to Rail+PRT, Source: (Muller, 2010)*

## Personal Rapid Transit Benefits

Personal Rapid Transit combines the positive aspects of the private car with the positive aspects of public transportation. The main benefits of a PRT systems are: Flexible Transportation (more than one route connecting an origin and a destination, which allows the service to run continuously even if a guideway is out of service), Economic Solution (less expensive per passenger kilometre than light rail or manual guideway transit systems), Safe Transportation (no traffic crossings, only low-speed merges and diverges), Secure Transportation (no crowds at stops and stations or in vehicles to eliminate terrorist attacks), Saves Time (no stopping at stations along the way (non-stop travel), less waiting time and less walking time), Sustainable Transportation System (less energy – uses less than one third of the energy per passenger kilometre of automobiles and public transit; less emissions, less noise and less vibration).

## Technical and technological characteristics of PRT systems

### Infrastructure design

This chapter presents the concept of the design and technology associated with PRT, which is basically a compilation of materials from various authors. Since this system is relatively new, and by that, unknown to the wider professional community and decision makers, the authors believe that this brief overview will be useful for future investors who do not participate in the development and research related to PRT systems at this time. There are three basic types of guidance in PRT: on open guideway; on captive bogey, and suspended, as illustrated in Fig. 2, 3 and 4.

*Fig. 2. Open guideway, Source: (Muller, 2007)*

*Fig. 3. Captive bogey, Source: (Muller, 2007)*

*Fig. 4. Suspended Source: (Muller, 2007)*

Each PRT system consists of automated vehicles and a supervisory (control) system and infrastructure. The vehicles (Transportation Pods or T-Pods) are small transportation units that circulate on light-weight infrastructure. Some T-Pods run on top of the guideway, while the others (bogeys) are held captive inside the guideway, and yet others run suspended from the guideway (which can make the switching from one guideway to another more complicated) (Muller, n.d.). Some systems follow magnets or wires buried in the guideway; some use on-board navigation systems backed up by sensors that confirm the T-Pod's position relative to objects along the guideway. Other systems can be physically constrained by wheels that follow guideway walls or other vertical elements.

The vehicles that are not in current use are maintained or they wait at stations or depots. This results in a reduction of unnecessary empty vehicle movements, which reduces energy consumption. The vehicles operate on the network moving along specially designed single-direction guideways that lie outside the set of interconnected loops. In this manner, it is possible to perform a multiplication of routes so that each source and destination can be connected. Each T-Pod can be guided individually through the optimal route towards its destination. An example of a PRT guideways is shown in Fig. 5.

The capacity of the guideway system in practice is between 1,000 and 14,000 passengers per hour (

Table 1) (Muller, 2011).

*Table 1. Maximum theoretical personal rapid transit capacities. Source: (Muller, 2011)*

| | **With Brick Wall Stopping (BWS)** | | | | | | | | **Without BWS** | | | |
|---|---|---|---|---|---|---|---|---|---|---|---|---|
| **Speed** | 24 kph | | | | 48 kph | | | | unlimited | | unlimited | |
| **Deceleration (G)** | 0.25 | | 0.50 | | 0.25 | | 0.50 | | 0.50 | 0.50 | 0.50 | 0.50 |
| **Minimum headway (sec)** | 2.0 | 2.0 | 1.4 | 1.4 | 3.2 | 3.2 | 1.7 | 1.7 | 1.0 | 1.0 | 0.5 | 0.5 |
| **Occupancy (passenger)** | 1 | 4 | 1 | 4 | 1 | 4 | 1 | 4 | 1 | 4 | 1 | 4 |
| **Passengers per hour** | 1,800 | 7200 | 2,571 | 10,285 | 1,125 | 4,500 | 2,117 | 8,470 | 3,600 | 14,400 | 7,200 | 28,800 |

If the vehicle occupancy is 4 passengers, and the headway is 2 seconds, then the capacity of the PRT system is 7,200 passengers per hour (number of passengers multiplied by 3,600 seconds divided by the headway). This is 92% of the capacity of the common LRT system (7,850 passengers per hour). The number of seconds between vehicles is called headway and it represents a critical factor for system capacity. Assuming that the average occupancy of private cars is between 1.1 and 1.5 passengers, the 2 second headway results in a capacity of 2,259 passengers per hour, which is almost identical to the capacity of one lane on a highway. As the costs of PRT systems are significantly lower compared to other modes of transport, it is usually much more useful to compare the costs for each mode of transport required to achieve the same capacity (PRT Consulting, 2008).

Public Rapid Transit stops are usually located at guideway intersections or between two intersections. The accessibility, presented in terms of walking time, depends on the distance between two stops, which is usually less than 500 m. The stops are positioned next to the travel line or beyond it (like a bypass), so the vehicles have to separate from the travel line in order to pick up passengers. In this manner, vehicles that transit by the stop do not have to wait for the dwelling vehicle to exchange passengers – they just continue driving towards their destinations without stopping. Places on the PRT stop where passengers alight and board (bays) can be positioned in these following configurations: in-line bays (Fig. 6), off-line bays with reversing (Fig. 7) and off-line bays without reversing (Fig. 8). The length of the platform depends on the maximum number of vehicles to be simultaneously accommodated for alighting and boarding.

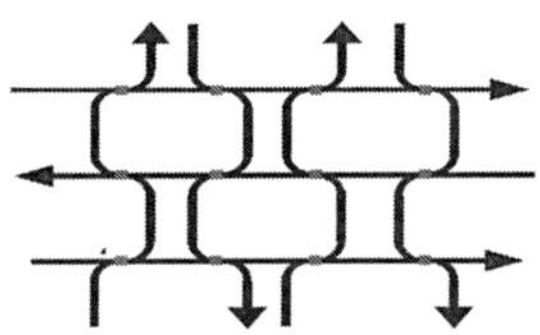

*Fig. 5. PRT guideways. Source: (Muller, 2003)*

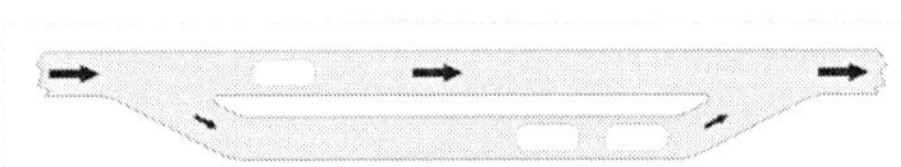

*Fig. 6. Offline station with in-line bays. Source: (Muller, 2009)*

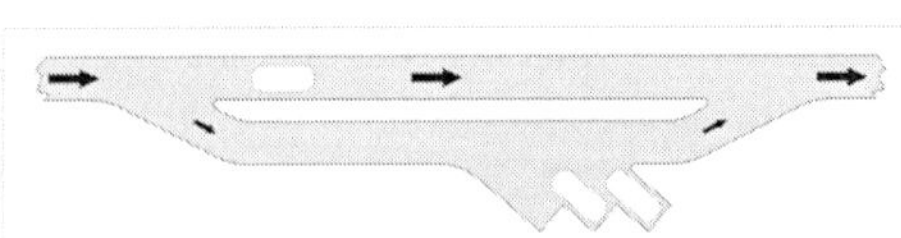

*Fig. 7. Offline station with off-line bays (requires reversing) (Muller, 2009).*

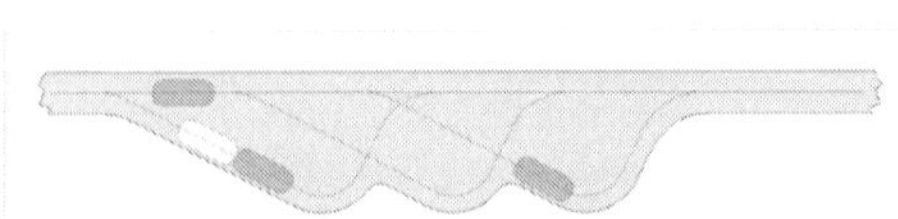

*Fig. 8. Offline station with off-line bays (no need for reversing) (Muller, 2009).*

## Navigation

The central computer is in charge of control, monitoring and optimisation of network traffic flows, interaction at the stops, alighting and boarding during dwelling time and passenger management in order to prevent crowding at points of passenger exchanges. Computer technology increases the reliability of the system, and special algorithms provide safer interactions between vehicles, therefore excluding human error. With the assistance of the Global Positioning System, the central computer can ensure real-time vehicle detection. The vehicles themselves are also equipped with odometers and gyro units supported by regular re-calibration to external reference points to make vehicle positioning even more accurate. In the event of an incident or malfunction, it is possible to re-route vehicles, and with the assistance of on-board CCTV cameras, video monitoring of the incident is also possible.

## Human machine interface (HMI)

Each vehicle is connected remotely to the central computer and also has a touch screen computer, microphone and speaker installed in it, as shown in Fig. 9.

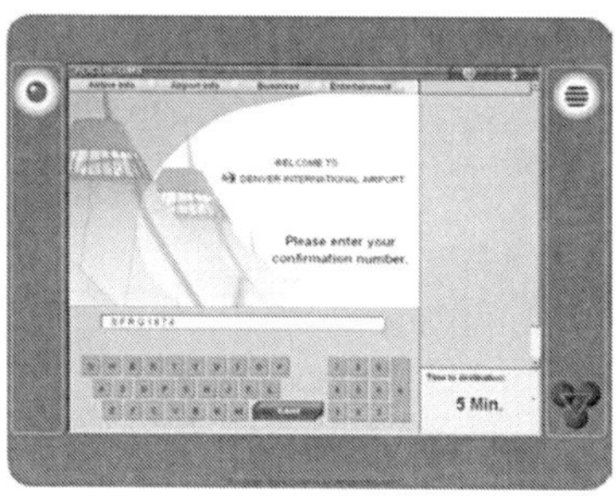


*Fig.9. Rendering of on-board touch-screen computer with CCTV, microphone and speaker. Source: (Muller, 2003)*

At every stop, passengers have access to an interface at which they can request a trip just by pressing a button. The choice of destination can be made at the stop or on the touch-screen in the vehicle. The fare payment can be done by card, and there are other methods of payment also available.

## Existing and planned networks

Table 2 shows two PRT networks, one quasi-PRT network in the operational stage and several in the planning stage.

*Table 2. Existing and planned networks*

| Location | Status | System | Date | Guideway | Stations / vehicles | Notes |
|---|---|---|---|---|---|---|
| Morgantown, West Virginia, US | Operational | WVU PRT | 1975 | 13.2 km (8.2 mi) | 5 / 73 | Up to 20 passengers per vehicle; some rides are not non-stop during low usage periods. |
| Masdar City, Abu Dhabi, UAE | Operational | 2getthere | 2010 | 1.5 km (0.9 mi) | 2 passenger, 3 freight / 10 passenger, 3 freight | Initial plans called for automobiles to be banned, with PRT as the only powered intra-city transport (along with an inter-city light rail). Due to costs, the initial plan was expanded to include electric cars and buses. |
| London Heathrow Airport, England | Operational | ULTra | 2011 | 3.8 km (2.4 mi) | 3 / 21 | In 2011, the PRT connected the terminal with a car park, and later the extension of the system is planned for the entire airport. |
| Suncheon, South Korea | Construction complete, operational testing | Vectus | 2013 | 5.3 km (3.3 mi) | 2 / 40 | The PRT connects Suncheon Garden Expo Korea to the Buffer Area station. |
| Amritsar, Punjab, India | Planned | ULTra | 2015 | 3.3 km (2.1 mi) | 7 / 200 | Construction has not started yet and reports indicate that PRT may be abandoned due to the development of bus rapid transit. |

*Source: (West Virginia University, n.d.); ("Personal Rapid Transit," n.d.); ("Ultra Global PRT," n.d.); ("Suncheon PRT," n.d.); ("VECTUS — Intelligent People Movers," n.d.)*

## Conclusions

Since 2001, the European Union has been funding the projects within the FP program. The goal of the program is to create a sustainable transport system at the social, economic and environmental scale. Due to energy consumption, environmental pollution, accidents, construction costs, a large number of empty runs, long waiting times and long driving times, conventional transport is not a sustainable urban mobility solution. At the present time, many solutions to the problems stated above are being considered because none of the solutions are effective enough just by themselves. At the same time, the appearance of PRT systems has provided enough encouragement to believe that public transport as well as transport between custom locations can be significantly improved. Rail and air transport are good solutions for long distances, but they have accessibility problems at the beginning and end stages of the trip. The congestion typical to city centres and transfers in public transport lead to delays and travel time extensions. Public Rapid Transit can overcome the problems related to accessibility, transfers and driving time extensions in densely populated central parts of cities. Since the PRT industry has not been fully established yet, steps for the proper evaluation of this technology have not been taken, setting up limits in the implementation of this technology. Therefore, these systems can only be applied with limited coverage, in specific areas and as feeder services to existing transportation systems for the time being. From the technological perspective, it is possible to implement the system in broader city areas. However, due to changes in a city's appearance and privacy intrusions, the technology is more likely to be applied only in specific parts of cities.

## References

Advanced Transit Association – ATRA. 2003. *Personal Automated Transportation: Status and Potential of Personal Rapid Transit Technology Evaluation.*

Irving, Jack. H.; Bernstein, Harry; Olson, C.L.; Buyan, Jon. 1978. *Fundamentals of Personal Rapid Transit.* Toronto: Lexington Books.

Mogge, John. 2009. *Automotive Industry Aviation Industry Rail & Transit Industry Built & Natural Environment Basic Technical Building Blocks for Personal Rapid Transit (PRT).*

PRT Consulting. *Improving Transportation through Innovative Engineering – PRT BENEFITS.* http://prtconsulting.com/docs/Benefits.pdf

Muller, Peter J. 2003. *Reinventing Airport Using Personalized Rapid Transport Pods (T-Pods).* Lakewood. http://prtconsulting.com/docs/Reinventing Airports.pdf

Muller, Peter J. 2007. "A Personal Rapid Transit/Airport Automated People Mover Comparison." Paper presented at the 9th International Air Transport Conference, Irving, Texas, August 19-22.

Muller, Peter J. 2009. "Open-Guideway Personal Rapid Transit Station Options". Paper presented at 12th International Conference of Automated People Movers, Atlanta, Georgia, May 31-June 3.

Muller, Peter J. 2010. *Improving Transportation Requires a New Solution.* http://prtconsulting.com/docs/MullerImprovingTransportationRequiresNewSolution.pdf

Muller, Peter J. 2011. *Personal Rapid Transit Capacity.* http://prtconsulting.com/docs/MullerPersonalRapidTransitCapacity.pdf

Google. 2014. "Personal Rapid Transit". Last modified February '09. http://www.2getthere.eu/?page_id=58

PRT Consulting. 2008. *Analysis of Capital Costs.* http://prtconsulting.com/docs/CapitalCosts.pdf

Google. 2014. "Suncheon PRT". Last modified February '09. http://suncheonprt.com/eco/index.php

Google. 2014. "Ultra Global PRT". Last modified February '09. http://www.ultraglobalprt.com/

Google. 2014. "VECTUS – Intelligent People Movers". Last modified February '09. http://www.vectusprt.com/EN/

Google. 2014. "Personal Rapid Transit (PRT)". Last modified February '09. https://transportation.wvu.edu/prt

# CHAPTER TWENTY THREE

# FLIGHTPATH TO AN ENVIRONMENTAL FRIENDLY AIR TRANSPORT

IVANA ČAVKA[1] AND OLJA ČOKORILO[2]

**Abstract** Environmental protection and safety are the main issues to be addressed with regard to aircraft operations. Since total $CO_2$ aviation emissions account for approximately 2% of global greenhouse gas emissions and the amount of $CO_2$ emissions from aviation is expected to grow by around 3-4% per year, European organizations have developed a range of standards and recommended practices to address aircraft noise and engine emissions, embracing technological improvements, operating procedures, proper organization of air traffic, appropriate airport and land-use planning, and the use of market-based options. This paper is based on contemporary environmental software tools for emissions evaluation in order to provide reduction and mitigation strategies at the national level. Therefore, Air Serbia's fleet was used to calculate emissions and devise medium-term mitigation strategies to improve $CO_2$ emissions in the aviation sector. The presented research is based on environmental trends in aviation projected to 2050, provided by ICAO.

**Keywords:** Aviation, Aircraft operations, Environmental protection, Aircraft engine emissions, Environmental tools

## Introduction

Effective and efficient performance of the principal mission of the aviation industry requires the balancing of numerous challenges, particularly with

[1] MSc Ivana Čavka, University of Belgrade, Faculty of Transport and Traffic Engineering, Vojvode Stepe 305, 11000 Belgrade, Serbia

[2] Assistant Professor Dr Olja Čokorilo, University of Belgrade, Faculty of Transport and Traffic Engineering, Vojvode Stepe 305, 11000 Belgrade, Serbia

regard to technical, economic and environmental issues, with safety always remaining of paramount importance. As air traffic is expected to double by the year 2030 (going from 5 billion to more than 13 billion in revenue kilometers per passenger, at an average annual growth rate of 4.9%) with continued growth to 2050 (at an average of 4% per annum), it is estimated that the consequent environmental footprints of aviation – aircraft noise, and emissions that affect local air quality and global climate – are also expected to increase, but at a rate slower than aviation demand. There is a significant commitment from all research stakeholders (e.g. manufacturers, research establishments, academia and government agencies) to investigate and develop novel technology and policy solutions aimed at reducing aircraft noise and engine emissions (Pham et al., 2010; Macintosh et al., 2009; Owen et al., 2010). In this regard, the Kyoto Protocol of the United Nations Framework Convention on Climate Change (UNFCCC) is of particular importance. The Protocol, which entered into force on 16 February 2005, requires the countries listed in Annex I of the Convention to reduce their collective emissions of six greenhouse gases, the most relevant one to aviation being carbon dioxide. The civil air transport sector analysis projects continuous growth in the next decade, with significant environmental consequences in terms of air pollution and climate change. The increasing public awareness of future climate change produces different measures for the civil air transport sector with regard to pollution mitigation (Döpelheuer, 2000; Mirosavljević et al., 2010b). Contemporary studies are based on developing an air pollution model aiming to use pollution measuring and monitoring points based on real air traffic (Mirosavljević et al., 2009; Čokorilo et al., 2010; Mirosavljević et al., 2010a; Mirosavljević et al., 2011a; Mirosavljević et al., 2011b). In some studies (Mihetec et al., 2011), $CO_2$ emission is used as environmental indicator for the evaluation of flight efficiency.

## Environmental Trends in Aviation

In order to achieve equilibrium between future commercial air traffic growth and environmental impacts, ICAO is taking concrete steps toward coordinated and comprehensive action to minimize aviation's effects on the environment. More precisely, ICAO has established three main environmental goals through its technical Committee on Aviation Environmental Protection (CAEP), whose principal objectives are to limit or reduce: 1) The number of people affected by significant aircraft noise; 2) The adverse impact of aviation emissions on local air quality; 3) The impact of aviation greenhouse gas emissions on the global climate. Every mentioned technical proposal developed by CAEP is analyzed according to four criteria: technical feasibility, environmental benefit, economic reasonableness, and the interrelationship among measures (ICAO Environmental Report, 2013). These goals are estimated in each three-year work cycle by using the latest input data and related assumptions, on the basis of which the present and future impacts and trends of aircraft noise and engine emissions can be assessed.

## Noise trends in aviation

The noise emanating from aircraft operations is the most significant cause of adverse community reactions that are related to the expansion of airports. Therefore, one of ICAO's main priorities is limiting or reducing the number of people affected by aircraft noise. The implementation of improved and advanced aircraft technologies and different operational procedures in the vicinity of airports are expected to reduce the amount of the total global population which is exposed to aircraft noise (Fig. 33-1).

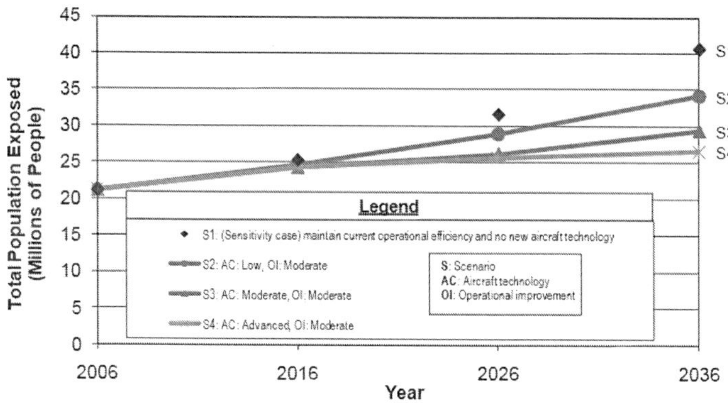


***Fig. 33-1:****Total global population exposed to aircraft noise above 55 DNL.Source: ICAO Environmental Report (2013)*

## Local air quality trends in aviation

Another public concern in the vicinity of airports focuses on the effects of aircraft engine emissions released below 3,000 feet and emissions from airport sources, such as airport traffic, ground service equipment, and de-icing operations. ICAO continues to develop measures aimed at mitigating the impact of aviation on local air quality. According to the current ICAO standards for emissions certification of aircraft engines, it must be demonstrated that the characteristic emissions of the engine type for HC, CO, $NO_x$ and smoke are below the limits defined by ICAO.

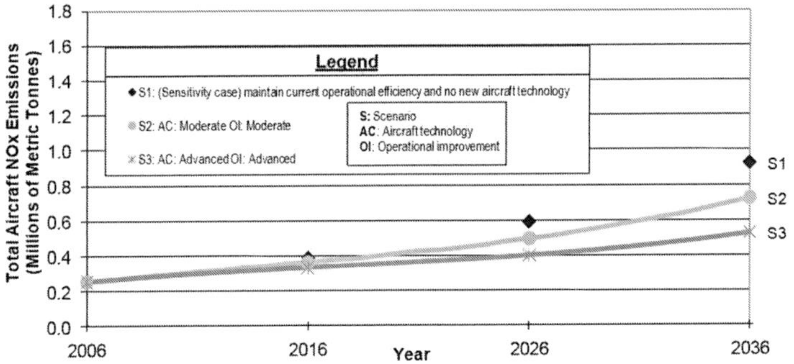


*Fig. 33-2.Total global aircraft $NO_x$ below 3,000 feet AGL*
*Source: ICAO Environmental Report (2013)*

The regulatory framework and drivers for local air quality measures, emissions inventories, emissions' temporal and spatial distribution, dispersion modelling, airport measurements, mitigation options, and interrelationships associated with methods for mitigating environmental impacts in the vicinity of the airport are documented within the ICAO Airport Air Quality Manual (Doc 9889) developed by CAEP in 2011. Dealing with the trends of local air quality, a number of scenarios have also been developed to demonstrate the reduction of aviation's impacts on the environment based on contributions from technology and operational improvements (Fig. 33-2).

## Greenhouse gas trends in aviation

$CO_2$ emissions from aviation have the most important influence on global climate because of its long atmospheric lifetime. From a macro perspective, the global aviation sector currently accounts for about 2% of the world's $CO_2$ emissions. For this reason, the most challenging task being undertaken by CAEP is the development of $CO_2$ emissions certification standard for aircrafts. CAEP also undertook the estimation of current and future progress toward the achievement of global environmental goals considering existing $CO_2$ trends. As depicted in Fig. 33-3, it includes various categories of mitigation measures: aircraft-related technology development, improved air traffic management and infrastructure use, and alternative fuels, assuming that 1 kg of jet fuel that is burned generates 3.16 kg of $CO_2$.

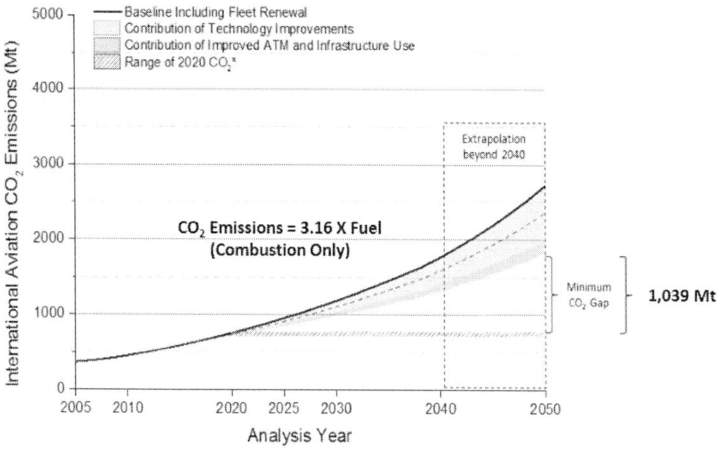


*Fig. 33-3.* *$CO_2$ emissions trends from international aviation (2005-2050) Source: ICAO Environmental Report (2013)*

## Emissions Inventory Evaluation – Case Study: Belgrade Airport

Serbia has a perfect geographical position, with the crossing of road and railway corridor 10 and the Danube River, corridor 7. Considering that transportation is a significant driving force of economic development in each country and that the precondition for this is the provision infrastructure, with passenger traffic being dominant in our country, it is necessary to point to the importance of air and water traffic, and their networking with road and railway traffic (Tomić-Petrović, 2012). Aviation-related activities are sources of a variety of air pollutants that can have potentially negative impacts on human health and the environment. However, the total amount of the relevant pollutants from a wide assortment and number of emissions sources that form an emissions inventory within the context of airport air quality can be generally classified into the following emission categories: $NO_x$ (Nitrogen oxides, including nitrogen dioxide and nitrogen oxide); VOC (Volatile organic compounds, including non-methane hydrocarbons); CO (Carbon monoxide); $CO_2$ (Carbon dioxide); PM (Particulate matter); and $SO_x$ (Sulphur oxides). Depending on the required level of accuracy and confidence in the results and the availability of supporting data, an emissions inventory at a given airport can be evaluated at three different levels of complexity, i.e. by the Simple, Advanced or Sophisticated Approach (ICAO Doc 9889, 2011).

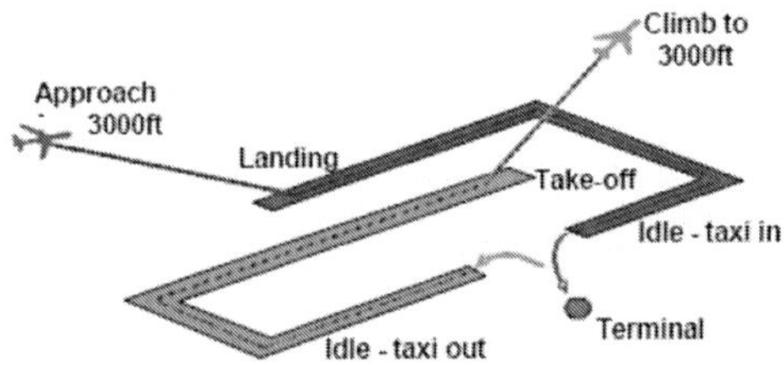


***Fig. 33-4.*** *Representative LTO cycle for aircraft engine certification*

In this paper, a Simple Approach has been used to estimate aircraft emissions using information from the dominant airport-related source during one day of aircraft activities at Belgrade airport. The methods proposed by ICAO for aircraft engine emissions calculations are functions of the four primary parameters (fleet mix, time-in-mode, main engine emission indices, and number of movements). The Simple Approach is only based on data concerning the aircraft fleet, emission index, and operations, whilst engine-specific emissions data have already been incorporated into the emission index and imply the use of the most common types of engine in operation internationally for a given aircraft type. Aircraft activity at an airport is quantified in terms of either landing and takeoff (LTO) cycles or operations. An operation represents either a landing or a takeoff, whereas two operations can equal one LTO cycle. ICAO has defined a specific reference LTO cycle for engine emissions certification purposes, and therefore it is used as a relevant element for the analysis. It relates to pollutants emitted in the vicinity of airports affecting local air quality below a height of 915 meters (3,000 feet) and contains assumptions for the time spent in each of the following conditions: taxiing, takeoff, approach, and idle (Fig. 33-4). The other sources of emissions that can be found at the airport (aircraft handling emissions, infrastructure or stationary related sources, and vehicle traffic sources) were not considered for the analysis. Summation of the total mass of emissions from each of these sources would be needed to obtain the emission inventory for the entire airport. Table 33-1 summarizes the data regarding aircraft/engine combinations, i.e. the fleet mix that operated at the Belgrade airport on 1 February 2014, as well as the corresponding number of LTO cycles. Additional information in conjunction with emissions factors for five pollutant types for each of the listed aircraft, together with associated fuel flow rates, were obtained from the ICAO Engine Emissions Databank (EASA, 2013).

***Table 33-1.*** *Aircraft engine emissions*

| | | LTO Emissions factors/airplane (kg/LTO/aircraft) | | | | | Fuel consumption (kg/LTO/ aircraft) | Number of LTO cycles (1 Feb 2014) |
|---|---|---|---|---|---|---|---|---|
| **Aircraft type** | **Engine type** | **$CO_2$** | **HC** | **$NO_x$** | **CO** | **$SO_2$** | | |
| A319 | CFM56-5A5 | 1153.4 | 0.297 | 4.367 | 3.175 | 0.365 | 365 | 20 |
| B737-300 | CFM56-3-B1 | 1238.72 | 0.418 | 3.595 | 6.517 | 0.39 | 392 | 11 |
| ATR72 | PW127F | 310 | 0.145 | 0.91 | 1.165 | 0.1 | 100 | 9 |
| Embraer 195 | CF34-10E5 | 957.48 | 0.848 | 2.836 | 6.761 | 0.155 | 303 | 4 |
| B737-800 | CFM56-7B26 | 1393.56 | 0.361 | 6.149 | 3.533 | 0.44 | 441 | 2 |
| Fokker 100 | TAY Mk650-15 | 1194.48 | 0.717 | 2.875 | 6.921 | 0.38 | 378 | 1 |
| A320 | CFM56-5-A1 | 1216.6 | 0.285 | 4.506 | 3.093 | 0.385 | 385 | 9 |
| Embraer 175 | CF34-8E5 | 761.56 | 0.018 | 2.222 | 2.065 | 0.155 | 241 | 1 |
| Embraer 190 | CF34-10E5 | 957.48 | 0.848 | 2.836 | 6.761 | 0.155 | 303 | 2 |
| Bombardier Q400 | PW120 | 320 | 0 | 0.755 | 1.12 | 0.1 | 100 | 1 |

According to the Simple Approach, aircraft engine emissions can be estimated for all relevant emissions types ($CO_2$, HC, $NO_x$, CO and $SO_2$) using the following methodology proposed by ICAO: For each aircraft type, the number of LTO cycles of that aircraft (over the assessment period) should be multiplied by the corresponding emissions factor for each of the pollutant types and then the values for all the aircraft should be added to obtain the amount of total emissions (in kg) for each pollutant (ICAO Doc 9889, 2011):

$$E_X = \sum_{All\ Aircraft} LTO_{Aircraft_Y} \times EF_X$$

where:

$E_X$ – Emission of species $X$ (in kg);
$LTO_{AircraftY}$ – Number of LTO (landing and takeoff) cycles of aircraft $Y$;
$EF_X$ – Emissions factor for species $X$.

The results of applying the abovementioned equation to the considered case of "Nikola Tesla" Belgrade airport on a given day are presented in Fig. 33-5 and Fig. 33-6. These figures show the distribution of the emissions values for each of the pollutant categories by aircraft type and the total values for each of the pollutant categories emitted during one day of operations at Belgrade airport, respectively. As can be seen from these figures, engine emissions from aircraft Airbus A319 mostly influence the local air quality in the vicinity of the airport. However, this result was expected, bearing in mind the dominance of this aircraft type at the airport, which is also a novelty in the fleet of the national airline.

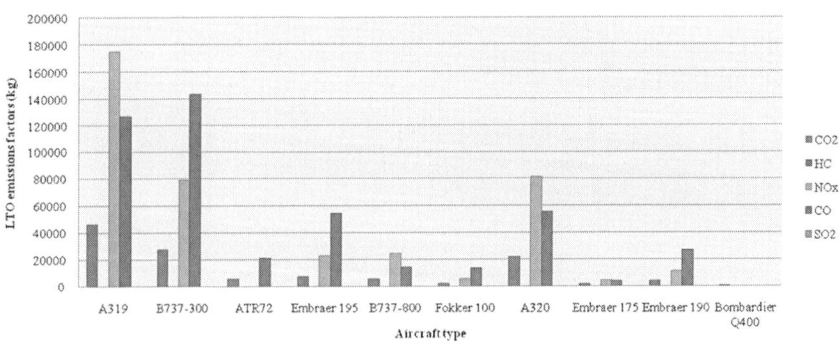


*Fig. 33-5. Distribution of the LTO emissions factors by aircraft type*

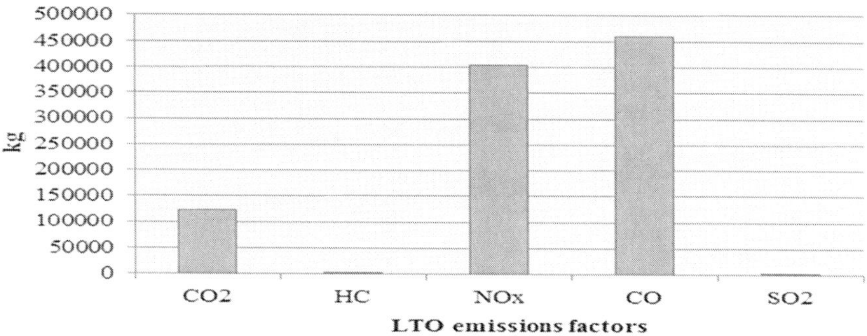


***Fig. 33-6.*** *Total LTO emissions factors generated during one day aircraft activities*

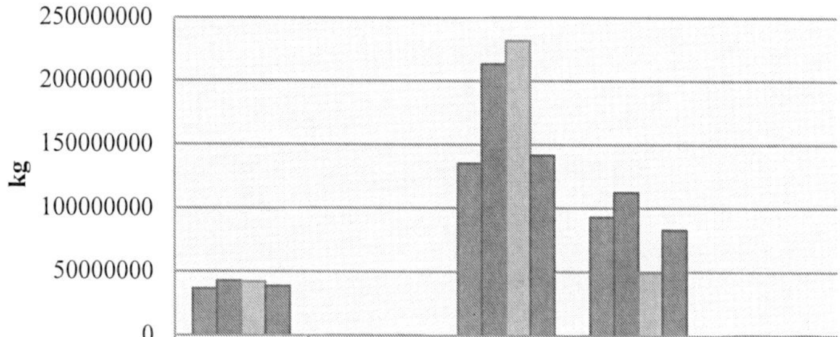


| | CO2 | HC | NOx | CO | SO2 |
|---|---|---|---|---|---|
| CFM56-5-A1 | 36498000 | 8550 | 135180000 | 92790000 | 10950 |
| CFM56-5B1/P | 42470400 | 21300 | 213360000 | 112140000 | 10950 |
| V2500-A1 | 41901600 | 2190 | 231480000 | 49680000 | 10950 |
| V2522-A5 | 38583600 | 900 | 141600000 | 82800000 | 10950 |

***Fig. 33-7.*** *Comparison of engine types in terms of emissions*

In this regard, it is important to examine the potential benefits of using a specific type of engine on this aircraft type and assess whether it would thereby be possible to achieve greater efficiency and lower emissions. The presented methodology of the Simple Approach was also used for this evaluation. Under the assumption of 15,000 LTO cycles and taking into account the four types of engines available in commercial use for aircraft A319 (two types of the CFM International Engines – CFM56-5-A1 and CFM56-5B1/P – and two types of the International Aero Engines – V2500-A1 and V2522-A5), it can be concluded that CFM56-5-A1 is the most efficient type of engine in terms of air pollutant emissions (Fig. 33-7).

## Conclusions

The trend of air traffic growth stipulates that international aviation organizations need to to develop policies, standards, guidelines and tools that will stabilize the global $CO_2$ emissions levels predicetd for 2020 by achieving a 2% annual improvement in fuel efficiency. A key factor that must be addressed when looking for ways to reduce fuel burn is the mass of the aircraft. However, the forecasted aircraft technology and operational improvements are expected to only partially offset the growth of $CO_2$ aviation emissions. Additional medium-term mitigation for $CO_2$ emissions from the aviation sector can potentially come from improved fuel efficiency. Alternative fuels may have the potential to make a contribution to the remaining gap, but their availability and potential contribution are still difficult to predict.

## References

Čokorilo, Olja. Gvozdenović, Slobodan. Vasov, Ljubiša and Mirosavljević, Petar. 2010. "Impact of Aircraft Emissions on the Environment." *Journal of Applied Engineering Science*, 8(3): 123-138.

Döpelheuer, Andreas. 2000. "Aircraft emission parameter modelling." *Air & Space Europe*, 2(3): 34-37.

EASA. 2013. ICAO Aircraft Engine Emissions Databank. Last modified April 2013. Accessed February 1, 2014.
http://www.easa.europa.eu/environment/edb/aircraft-engine-emissions.php/.

ICAO Doc 9889. 2011. *Airport Air Quality Manual.*

ICAO Environmental Report. 2013. *Aviation and Climate Change.*

Macintosh, Andrew and Lailey Wallace. 2009. "International aviation emissions to 2025: Can emissions be stabilised without restricting demand?" *Energy Policy*, 37(1): 264-273.

Mihetec, Tomislav. Odić, Denis and Steiner, Sanja. 2011. "Evaluation of Night Route Network on Flight Efficiency in Europe." *International Journal for Traffic and Transport Engineering*, 1(3): 132-141.

Mirosavljević, Petar. Gvozdenović, Slobodan and Čokorilo, Olja. 2009. "Cost of Fuel Consumption Increase to Airline: an Assesment of the ACARE Pollution Standards." In *Proceedings of 13th Annual World Conference, Air Transport Research Society (ATRS) World Conference*, Abu Dhabi, 27-30 June, pp. 1-10.

Mirosavljević, Petar. Gvozdenović, Slobodan and Čokorilo, Olja. 2010a. "The Turbo Fan Transport Aircraft Air Pollution Emission Footprint." In *Proceedings of 13th International Conference on Transport Science ICTS 2010*, Portorož, Slovenia, 27-28 May, pp. 1-10.

Mirosavljević, Petar. Gvozdenović, Slobodan and Čokorilo, Olja. 2010b. "The Transport Aircraft Pollution Cost Reduction Strategy." *FME Transactions*, 38(4): 157-166.

Mirosavljević, Petar. Gvozdenović, Slobodan and Čokorilo, Olja. 2011a. "A Model of Air Traffic Assignment as Part of Airport Air Pollution Management System." *Aviation*, 15(4): 92-100.

Mirosavljević, Petar. Gvozdenović, Slobodan. Čokorilo, Olja and Vasov, Ljubiša. 2011b. "The Turbofan Transport Aircraft Pollution Calculation Software." In *Proceedings of the International Conference on Climate Friendly Transport*, Belgrade, Serbia, 16-17 May, pp. 379-396.

Owen, Bethan. David S. Lee and Ling Lim. 2010. "Flying into the future: aviation emissions scenarios to 2050." *Environmental science & technology*, 44(7): 2255-2260.

Pham, Viet Van. Jiangjun Tang. Sameer Alam. Chris Lokan and Hussein A. Abbass. 2010. "Aviation emission inventory development and analysis." *Environmental Modelling & Software*, 25(12): 1738-1753.

Tomić-Petrović, Nataša. 2012. "Transportation and Environmental Protection – Challenge for the Future." In *Proceedings of International Conference on Traffic and Transport Engineering*, Belgrade, 29-30 November, pp. 461-464.